UNDERSTANDING HF WIRE ANTENNAS
WRITTEN FOR BEGINNERS AND AS A REFRESHER

BY JERRY E. BUSTIN

6TH REVISION

Copyright © 2020 by Jerry E. Bustin
All rights reserved. This book or any portion thereof may not be reproduced or used in any manner whatsoever without the express written permission of the publisher except for the use of brief quotations in a book review.

Edited by Scott Miller, KD7YDQ; Glynn Hughes. VK5GP; and Professor Dan Cadwell, Boise State University
Proofread by Kristen Corrects, Inc.
Cover art design by Steven Spence KF6TSD

First edition published 2020

UNDERSTANDING HF WIRE ANTENNAS
WRITTEN FOR BEGINNERS AND AS A REFRESHER

JERRY E. BUSTIN

TABLE OF CONTENTS

Introduction: The Misunderstanding of Isotropic, Dipole, and Gain i
Chapter 1: Undaerstanding Wire Antennas 1
Chapter 2: Quarter-Wave Ground Planes 7
Chapter 3: Real True Gain or Isotropic Gain 14
Chapter 4: Antenna Gain in Real Time 16
Chapter 5: Bandwidth 20
Chapter 6: Reciprocal Relation between Tx/Rx 23
Chapter 7: Easy up Antennas 27
Chapter 8: Basic MonsterAntennas 35
Chapter 9: Construction of Antennas 48
Chapter 10: Termination Resistor 56
Chapter 11: Finding the Velocity K-Factor 60
Chapter 12: Antenna Tuners 65
Notes 68
Author's Thoughts on Gain 70

Introduction

The Misunderstanding of Isotropic, Dipole, and Gain

All the different views and explanations concerning isotropic and dipole antennas have created quite a controversy over the years. However, understanding the true gain of a dipole antenna will help clear up any misconceptions.

Over the last few decades, the author has led many training classes that have assisted graduate students in correcting their understanding of antenna gain. He authored this book to partner his knowledge with other advanced engineers, and to assist the beginner with basic understanding. It is his desire that in the future, universities will base their hiring on the knowledge possessed by the candidate, not on their Ph.D. status. This would greatly improve the written word on antenna construction. In addition, it would bring back the analytical and applied knowledge missing in today's university classroom.

There seem to be two separate thoughts among groups of trainers and ham operators concerning antenna gain. One group believes that reference antennas are isotropic radiators. The other group understands that this is a manufactured idea. The author does not mind the isotropic radiator's misconception; he just does not appreciate reading this notion in a training manual on sale in a university bookstore.

In quoting the author, "Years ago I met a few ham operators that contacted me concerning antenna construction. They had ideas on the gain of a dipole, contrary to the lecture I had presented. I asked for their sources, and they stated ham programmers wrote the software. I didn't want to debate these sincere cohorts, but I did mention the fact that the software is only as good as the person writing it." The author continued by telling me to arm myself with as much knowledge as possible. That alone can keep you from buying snake oil.

Jerry E. Bustin's ability to explain technical subjects to novice and journeyman alike is unmatched. Making difficult subjects seem simple is a talent few people possess. However, when your heart's in what you teach, it is possible, as the author has demonstrated. While our nation's colleges have reduced hands-on training and increased online computer classes of theoretical delivery, this author insists on application and analysis of all technical subjects.

This book concerning HF wire antenna construction gives step-by-step beginning instructions of antennas—not only building antennas, but also the understanding of what works well and things that can go wrong. The reader has the opportunity to not only learn how to manufacture their own product, but also to analyze the quality of every aspect of the device. Arming the ham operator with the ability to construct antennas enhances their understanding to the point of quality ham operations.

In the end, gaining the information in this manual should equal a more quality time in ham operation.

Dan E. Cadwell
Professor Emeritus, Boise State University Who's Who Among American Teachers

Chapter 1

Understanding Wire Antennas

This section is for beginners or as a refresher. It is understood that you, the reader, have some brief knowledge of a dipole and ham radio. This is based on the fact that someone led you to this book or the store where it is sold. With that said, I will try to explain the basic wire antenna concept.

Any antenna must have a smooth transfer of power from one transmission medium such as a circuit, introduced into a wire antenna to another medium—surrounding space. On the wire side of the circuit, the usual wire network conditions of impedance match and inherent low loss must be satisfied. If you do not understand this, it will become clear in the following chapters.

The antenna must launch the energy (radio waves) into space with little loss of power—and usually with the added requirement of the energy in the proper directions. Essentially, the antenna is a wire circuit termination of special size and shape, so placed to perform these functions efficiently.

A short length of wire energized with a high-frequency voltage is a simple form of antenna. The current flowing along the antenna length is the source of the radiation with the moving electrons producing the radio wave, or field reaching out to great distances.

Any wire, when exposed to a radio wave, has a voltage induced in it and produces a potential (current) across an associated resistance of that wire. Therefore, the wire delivers the energy collected from space to a conventional radio receiver. Any wire, whether it's electrical wire, barbed wire, or fence wire, is a potential antenna in waiting.

Each of the differing types of antennas will produce a different type of a radiated pattern. The distribution of the radiated field in space depends upon the current distribution in the antenna and its aspects at various angles from the antenna. A short but straight energized antenna wire (conductor) has no current flow at the ends of the wire, when viewed from the ends of the wire. Therefore, the (current) radiated field at either end is essentially zero. At a progressively greater angle from the antenna wire and around to a line perpendicular to the antenna wire, the current appears in greater perspective so that the intensity of the radiated field becomes increasingly stronger in the center of each quarter-wave point. On the other hand, voltage is at a zero at the apex (center of a dipole) and highest at the ends of a dipole.

Pattern of Plane

Remember, an antenna transmitting in a vertical pattern will be displayed in a horizontal plane and a horizontal pattern will be displayed in a vertical plane. Many radio operators have a difficult time with this concept.

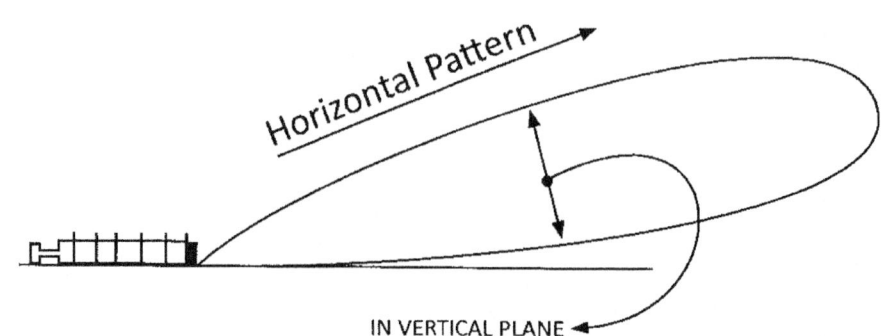

As you know from childhood, a wire can carry a lethal current and voltage if wired into a proper circuit with a high voltage load. We learned as children not to touch or stick something metallic into a common house outlet. It could kill you. This is because metal is a conductor of electricity and *CAN CARRY A LETHAL CURRENT*.

Matter is either a conductor or an insulator (yes, it is possible to have both, but it will not be discussed in the book). Just understand a conductor (like copper wire) enables current to flow. An insulator (like glass) is designed to stop electrical current flow. Wire is a good conductor; however, it can have resistance to current if the wire has a small diameter running over a long distance.

Antennas are indeed conductors. Lengthening a straight conductor from exceedingly small dimensions up to slightly over a half-wave (of the frequency adjusting) will not change the radiation pattern much at all. However, the impedance is greatly affected. That is to say, the input voltage and current does vary. A short center-fed conductor will react as a large reactive component and will act much like that of a leaky capacitive element. Inductive reactance will increase in proportion to an increase in antenna wire length. This increase will eventually counteract all the capacitance, thus only a resistance appears at the feed point (apex, where the coax is connected to the antenna). The longer the wire, the more the resistance displayed in the antenna.

Resonance occurs when an isolated conductor is at or near the half-wave point for the frequency adjustment. In electricity, resonance is the condition by which the greatest amount of current will flow with less opposition to resistances of a certain frequency.

DO NOT FORGET THIS RULE: The thicker the antenna wire, the less the length will be in any given antenna!

Your overall length of your antenna will be longer if you use thin wire whereas the same antenna would be shorter if you use thicker wire and would become more broad-banded. A resonant condition in any given antenna is obtained at exactly one half-wavelength for an infinitely thin wire gauge. This occurs at lengths that are a few percent under a half-wavelength for conductor of practical thicknesses.

Remember: Resonance occurs at every odd multiple of half-wavelengths.

At the half-wave resonant point, in the basic dpiole antenna length, the input impedance of a very thin wire or thicker wire does not vary by much but are about 65 to 73 ohms in each case. Sloping a dipole or inverted vee at the far ends, below the height of the center apex feed point, produces lower impedance. Inverted dipole impedance is nearer to 50 ohms. Stranded wire displays a better-radiated RF field opposed to solid wire. Stranded wire has more surface area for greater skin affect. Any time you move the ends of a wire antenna up, down, or make it longer or shorter, it will affect the SWR readings.

To obtain a good SWR (standing wave ratio)—the lower the SWR the better—you must take into consideration the length of the center apex gap. The amount of gap spacing between each quarter-wave section at the apex is accountable within the total half-wave overall length.

Shorter straight wires (rigid) have progressively lower resistances (in ohmage), but they are always accompanied by a capacitive reactance (that can be seen in the SWR). This reactance should be tuned out by a capacitor. (Most hams do not do this and then wonder why their antenna does not display a true 1:1 SWR when tuned.) Simply connect a capacitor at the feed point, the center coax conductor (connected at the positive side of the antenna wire) to the coax ground braid (connected to the negative side of the antenna wire) radial quarter-wave vertical antenna or in a horizontal dipole antenna. SWR is affected by wire that stretches due to weather.

Most antennas work well with a .47pf/1000Volt capacitor. You may even try a ceramic variable .10 to .50 or even 10 to .125pf or larger. However, your antenna should work simply fine without one. It's your choice, and it's fun to experiment.

A half-wave folded dipole has an input resistance that is four times that of a dipole about 290 to 300 ohms. The radiation resistance varies with the position of the input feed point, wherever there is a change in the current amplitude along the length of the conductor.

Why does an antenna change SWR with the weather? Here is a good reason. If your antenna system changes SWR on some days—i.e., mornings or rainy days—it is due to the ground to air capacitance varying.

That's right! It is the capacitance in the air molecules with respect to static ground current. Vertical antennas are mostly affected. Changes in atmospheric conditions will have some effect on all antennas at one time or another; some are hardly noticeable while verticals are at the top of the list.

Type of terrain	Dielectric constant	Conductivity	Like
Sea water	80	5	Ocean
Fresh water	40	4	Lake
Moist wet soil	30	2	Swamp
Fertile soil	15	0.1	Good farmland
Rocky ground soil	7	0.005	Poor farmland
Rocky	5	0.001	Poor soil
Dry desert soil	4	0.01	Cactus country
Super dry desert soil	3	0.001	Death valley region

Depending on the ground potential where you live, you may not have noticed the changes. Saltwater is the best at 80 constants and dry soil the worst.

Thus, if the ground potential changes, so shall a vertical impedance change (fluctuate). This is more likely than that of a dipole or beam antenna system. Effects will most likely be higher in drier ground regions than the wetter areas of our country. If you live in a wetland like the state of Florida, will the reception be worse all the way around for one who lives in a hot, dry area such as Phoenix, Arizona? Yes indeed!

A *good grounding system* is paramount and the way to improve your system over neighboring hams. A single ground rod placed outside your shack is inadequate. (If you have no other choice, it will have to do.) However, it is ten times better than a water faucet! Your ground strap or braid out to your first ground rod should be as short as possible. No matter what your situation is, make your braid as short as possible and *without* any coils in your braid.

Using four rods, as illustrated below, in a six- to eight-foot square, connect all rods together. The more rods you have, the better. Place a copper wire mash grounded grid if you like. Search the web for RF grounding systems.

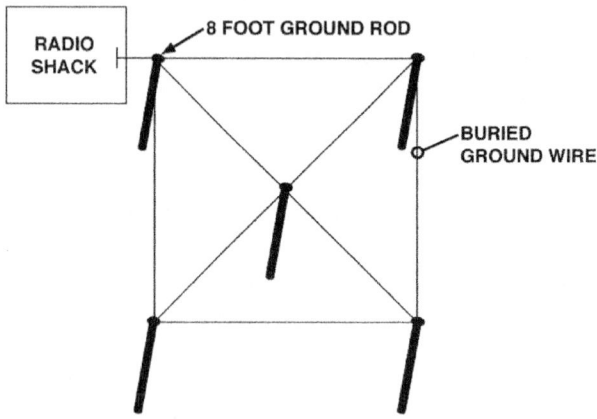

Drawing not to scale

The Q (or bandwidth) of any antenna is the measurement of the bandwidth of an antenna relative to the center frequency of the bandwidth.

Antennas with high-Q are narrow-banded; antennas with a low-Q are wideband. A mono band dipole will have a low-Q while a trap or coil dipole antenna will have a high-Q.

Chapter 2

Quarter-Wave Ground Planes

When you load up a short vertical wire with a transmitter and the wire is increased in length with respects to earth ground, it will attain a resonance at about a quarter-wavelength. This you can take to the bank! If you took a half-wave dipole and positioned it vertically to ground, then you would have a half-wave vertical dipole. If you remove the bottom (negative) side wire of the same dipole—that is to say the wire that the coax ground braid is attached to and replace it with earth ground or a car frame—then you will have a quarter-wave antenna, with half the potential of a dipole.

It now may be a vertical dipole half-wave cut in two by a ground plane at the center and the lower half removed. Therefore, a dipole is said to be two quarter-waves in phase, equaling a resonance half-wave dipole. If each of the quarter-waves in a dipole is called an element, then how on green earth does a single quarter-wave element standing alone resonate?

Simple: It does not. Read on.

Any quarter-wave antenna must have a half-wave or a potential of its missing half-wave in order to obtain resonance. What makes it work properly is the unseen lower half of the half-wave is often viewed as the image source. On the following page, please study the drawing. An image source can be anything metallic in nature or you may even place your newly designed vertical directly on the ground using the earth as the image source. "Hey! Man did not invent the nature of things. We only figured out how to use it and how it works." That is what the old timers said many years ago.

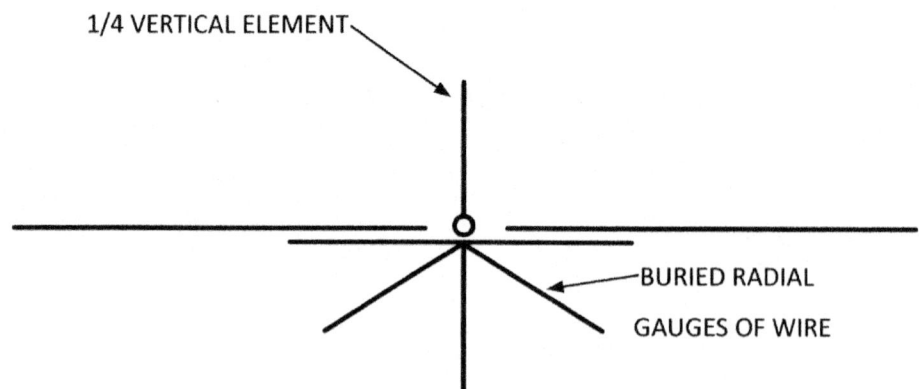

Electrical imaging – Mirror effect of a quarter-wave

The hidden unseen lower half (of the half-wave) is often viewed as an image source for the reflected-wave component of the propagated wave in question.

The radiation resistance of the quarter-wave section is about 16 to 36 ohms—about half of a dipole and over a perfectly 100 percent true conducting earth ground. Its radiation pattern is similar to that of the upper half of a dipole (73 ohms) placed in a vertical plane in free space. Therefore, a dipole uses two each quarter-wave sections, which creates a half-wave at 73.5 ohms. It does not take a rocket scientist to conclude that if you employ a single quarter-wave section in the use of a quarter-wave ground plane, the input resistance would be half of 73.5 ohms and not that of 52 ohms. It would be near 36.75 ohms. What we have here is a mismatch in impedance. The feed line coax is 52 ohms, and the vertical antenna is 36.75 ohms. How is this corrected? We must increase the ohmage of the antenna and bring it close to 52 ohms by adding radial-elements to the design of the quarter-wave vertical. With the imperfect earth ground, the distribution of the radiated energy is thus modified by the earth losses.

The radials are attached to the negative or grounded side of the coax ground braid. You will place five (5 each) wires or more and spread them out like spider legs in all direction from the bottom of the base of the vertical antenna. Each radial will be a quarter-wave plus .05 percent of one wavelength long. For those of you who do not know how to do the math, it will be explained in detail later in the book.

In the magnetic invisible electrical field radiated out into space by a quarter-wave antenna placed vertically over a ground plane, the radiated pattern of the magnetic RF antenna field in the region directly below the ground plane is eliminated (electrically induced); thus, RFI is at a null. The antenna power otherwise used to produce this lower half of the radiated energy in free space can now be added to that generating the field above the plane.

A dipole, if moved into close proximity with a perfect conducting ground plane, would be 5.6dB. This is better than, let's say, a standard quarter-wave HT whip (rubber duck) placed on the same ground plane in free space over an isotropic radiator or 3.5dB over the rubber duck if compared to a quarter-wave vertical placed in a vertical ground plane. Ground loss will be explained in a moment. However, now is as good of time to expose some of the author's frustrations! Informing you about dBi is a must.

An old-time ham operator told the author something he will never forget. He now lives by this statement. Here are his thoughts on dBi:

> "Some CEO and his cronies a long time ago were all upset because their antenna competitors were stating as a matter of fact that their antennas had more gain than their competitors! In addition, of course, as with all top company executives, they hired some firm to come up with a solution—

darned if they did not! As I live and breathe, the answer was so simple. It was so simple in fact that in a court of law, it was bulletproof!"

What was it, you might ask? The old timer continued:

"Manufacturers of antennas had to find a solution so their normal gain antennas would appear to the public as power gain grabbers! Otherwise hams and SWLs will only buy an antenna if it has the most gain!"

Meaning, before they came up with this new marketing scheme, reference antennas were either a half-wave dipole or a quarter-wave vertical. This is the way it should be represented. It's the honest way.

"Presto! The advertisers came up with a new all-imaginary reference antenna and called it the isotropic zero gain antenna! WOW!"

There you have it. Just beware and arm yourself with the old timer's information.

To limit ground losses, particularly in transmitting antennas, a copper counterpoise consisting of a grid of wire laid on the ground—or best, buried slightly under the ground—is preferred for true mirror vertical ground imaging.

Remember: This network of arranged grounded counterpoise is called radials. Radials are always in the ground (negative) setup. Radiators are always in a vertical (positive) setup, speaking in terms of vertical antennas. However, this does not apply for a J-pole antenna system, whether the match is with straight feed tuning or better yet with inductive tuning (coiled). All J-poles require no counterpoise radials.

Remember: When tuning any antenna for center band coverage:

(B) If your SWR is better at the low end of the band than at the high end of the band, *shorten it!*

(X) If your SWR is better at the high end of the band than at the lower end of the band, *lengthen it!*

Doing so will move the SWR curve in that direction.

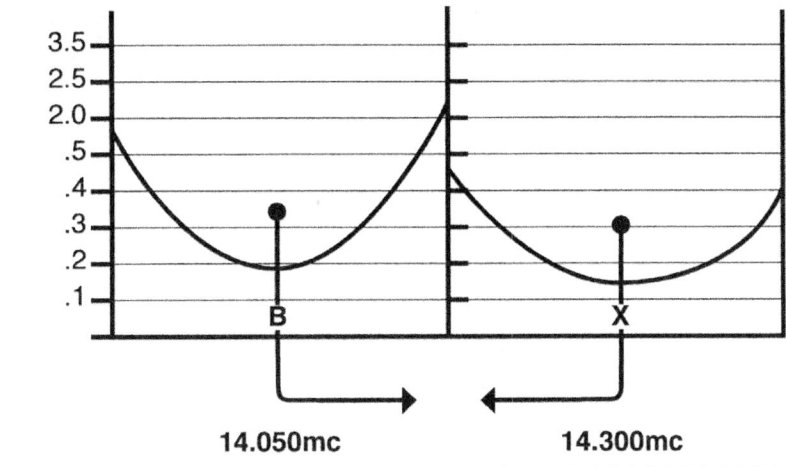

****

Resonant: Antenna designed for a single band.

Non-resonant: Antenna designed for more than one band.

Balun

A balun is a transformer. The balun will match your 50-ohm radio to your antenna system. The chart shows what value your balun must be. For example, if your antenna is a single-wire diamond rhombic with the impendence of 800-ohms you will be looking at a difference of 50/800 = 16:1 balun will be needed, as seen below.

BALUN RATIO

OHMS	Ratio	OHMs
50	1:1	50
50	2:1	100
50	3:1	150
50	4:1	200
50	5:1	250
50	6:1	300
50	7:1	350
50	8:1	400
50	9:1	450
50	10:1	500
50	11:1	550
50	12:1	600
50	13:1	650
50	14:1	700
50	15:1	750
50	16:1	800
50	17:1	850
50	18:1	900
50	19:1	950
50	20:1	1000

Remember: A current balun's sole use is to distribute current evenly on both sides of a balance antenna, such as a dipole, inverted-V, etc. A balun will not improve your SWR.

Balance: Means both sides of the wire antenna are the same length.

Unbalance: One side of the antenna wire is longer than the other side or using one side only as in a long wire, wave, and/or beverage, such as a Texas long-gun.

UNDERSTANDING HF WIRE ANTENNAS

Current balun should make life easier to match antennas to your 50-ohm radio or receiver.

An in-line isolator is a transformer, which minimizes or eliminates altogether unwanted RFI, TVI, or RF-current from running up and down your coax cable, better known as a choke.

Some abbreviations used in this text and drawing:

- dB: decibel
- HF: high-frequency
- MARS: Military Affiliated Radio Systems
- Mc: megacycle
- QRM: manmade interference noise
- QRN: atmospheric noise
- RF : radio frequency
- RFI: radio frequency interference
- SWL: short wave listener
- SWR: standing wave ratio
- TVI: television interference

Chapter 3

Real True Gain or Isotropic Gain

The author's opinion: "If I cannot see it, touch it, then I tend to question what they're selling concerning dBi." Try to be as realistic and non-biased as you can when someone speaks in terms of the gain dBi. If they rated it over a quarter-wave vertical or half-wave dipole then you can understand it to be true dB gain. Because you can see the dB on the S-meter on your radio, you can touch the reference antenna, and the results will not be from what a computer program says it should be.

Generally, the direction of maximum radiation from any antenna is the direction of principal concern in amateur radio communications. The power propagated in that direction is a direct measurement of beaming power of the antenna. A vertical antenna under any given test can be rated by comparison with a quarter-wave vertical ground plane because it presents a zero (0) gain dB or in with a half-wave dipole for the same reasons, or for an isotropic radiator. Most manufacturers will use a comparison to that of the isotropic radiator antenna, which they place at a zero gain reference. They list it as dBi. (It adds gain to theirs. Check their ads for yourself. You cannot prove something you cannot be seen!) They call it *theoretical free space.*

Let us come down to earth. So how can an average Joe QC public ham radio operator determine if their home-built or newly bought antenna has gain? It can be done and without those expensive computer programs and in-lab test sets. Your preference.

The propagated power in a given direction is a measure of the beaming power of the antenna under test, as mentioned earlier. The antenna is put to a comparison of its electromagnetic radiated lines of field—along this line with that of a basic or reference antenna of equal total radiated power at (1-watt). WOW! What a mouthful.

Simply put, the ratio of these two values—expressed as the square of the field intensities at a given point—is called the directivity gain of the tested antenna over the reference antenna or the difference of.

The reference antenna may be a theoretical point source, again called an isotropic radiator, which is assumed to radiate uniformly in all directions, up/down, left/right, and therefore has a spherical radiation pattern. The isotropic radiator has a given gain of pure unity and all other antennas will have a greater gain, since the slightest deviation in the spherical pattern has an increase in the field as in some direction over that for the uniform field. This is always compensated for by a smaller field in some other direction, since the total radiated power is assumed constant in nature.

In other words, if you have an antenna for which you would like to know just how much gain it really has, you must test it against another antenna—to compare the differences (side by side). It's called a reference antenna of a known value. A theoretical isotropic radiator (dBi) or a quarter-wave vertical/half-wave dipole (dB) should be used. This would be your choice.

Chapter 4

Antenna Gain in Real Time

Type of Antenna	Frequency	Size	Gain
Tower, flattop, inverted L, Inverted T, Lazy H, etc.	0.05 to 20 mc	Up to several hundred feet high	0 dB
Wave, beverage, Long Gun, Long wire	2 to 30 mc	10 to 100 feet long	0 to two digit dB
Horizontal rhombic	0.03 to 40 mc	3 half wave to miles long	6 to 18 dB
Horizontal dipole, folded dipole	2 to 500 mc	1 to 200 feet long	0 dB
Vertical dipole	3 to 500 mc	Normally 1 to 15 feet each side	0 dB
Quarter wave vertical	20 to 500 mc	1 to 30 feet high	0 dB
Mobile HF whip	1.8 to 500 mc	1 to 10 feet high	has a loss
Director - reflector	7 to 500 mc	1 to 20 feet square	6 dB
Dipole w/plane or corner reflector	30 to 1000 mc	1 to 20 feet square	8 to 18 dB
Parabolic reflector	1000 to 8000	1 to 10 feet diameter	15 to 45 dB
Simple horn	Above 1000	few inched to 3 feet	5 to 15 dB
Horn with lens	Above 1000	Several feet in cross section	20 to 40+ dB

Because of the size of larger HF antennas, for classroom purposes we will discuss the smaller 2-meter version.

Okay, you have spent a great deal of time inventing or building a 2-meter antenna and now you want to test it for use. Whatever it is, you now want to know how much gain it has. When finding gain of a new antenna, you should compare it with a reference antenna.

Remember: A quarter-wave ground plane has zero gain and can always be used as a good reference antenna in the field.

To perform this test in the field, you should have the following:

- One 12-volt battery. A car battery or deep cycle battery is fine.
- One folding table or suitable table.
- Two each, hand-held radios.
- Two test mobile radios that have the old analog needle S-Meter (one in the mobile and the other on the test table—set to 1-watt out).
- Two moveable quarter-wave ground planes and a stand (mount one at the table and the other is a mag-mount for the mobile).
- Two folding chairs.
- Coax jumper cables.
- Antenna analyzer.
- Drinks and food.

If possible, you should locate an area of land that is flat for about seven to ten miles in length. Drive out to this site with your partner. Now, set up the folding chairs and your worktable. Place all your equipment on the table. Turn on the radios. Set the mobile radios on a simplex channel not in use (to be used for the testing—for gain at one-watt output). You can choose another frequency for the hand-held radios for your communication between partners on another simplex channel as you test. Someone will stay at the test site while another one will leave down range in a vehicle with his or her own quarter-wave mag-mount vertical antenna (mounted on the top of the vehicle) connected to a test radio at 1-watt output.

At the test table site, you will have a quarter-wave ground plane mounted on a stand and set up as your reference antenna. You will also have your new vertical beam antenna. It is the vertical beam antenna for which you need to know the true gain. (Repeat the same process but with a dipole for a horizontal testing.)

As the vehicle drives away from you, ask your partner via your hand-held radio to transmit a one-watt continuous carrier from the mobile radio connected to the vehicle's quarter-wave ground plane. This should be on a mag-mount attached on the top of the roof. When your partner keys down, a one-watt signal will be transmitted continuously into the open air (in space) around you—striking your reference antenna, causing the S-meter's needle to swing forward on the meter to the strength the signal is receiving.

As the vehicle drives further away into the distance, the strength of the received signal striking the reference antenna (at the test table site) will weaken, thus causing the needle on your 2-meter radio's S-meter to lower in dB strength. As the needle lowers, tell your partner via your HT to stop the vehicle the moment you see the needle reach zero.

Now, both reference antennas are calibrated to a dead reckoning of absolute zero dB in real live gain at one-watt output. The driver will stop transmitting and will watch the S-meter on the mobile radio. It is at this point that the person at the table test site will remove the coax cable from the reference antenna and connect it to the new antenna to be tested. In this case, it is the new beam, aimed at your partner.

Let's for sport say the test antenna is a 4-element 2-meter quad, built for vertical. With your quad antenna ready and connected, and your partner way down range in a vehicle listening to you on an HT while looking at the S-meter on his mobile radio, you (at the test table site) will now transmit a 1-watt carrier into the quad.

Remember: Whatever your partner (in the vehicle down range) reads from his or her mobile's S-meter will be the true gain of the test.

QUAD over that of a quarter-wave reference antenna or a half-wave dipole. Let's say the mobile radio's S-meter jumps from 0dB to 6dB. The test quad has a true gain of 5dB over a quarter-wave antenna, confirmed!

Testing other newly constructed antennas this way, the process will work the same. That's it. That's how the old timers did it and how the military taught me.

YOU WILL HAVE FUN DOING THIS!

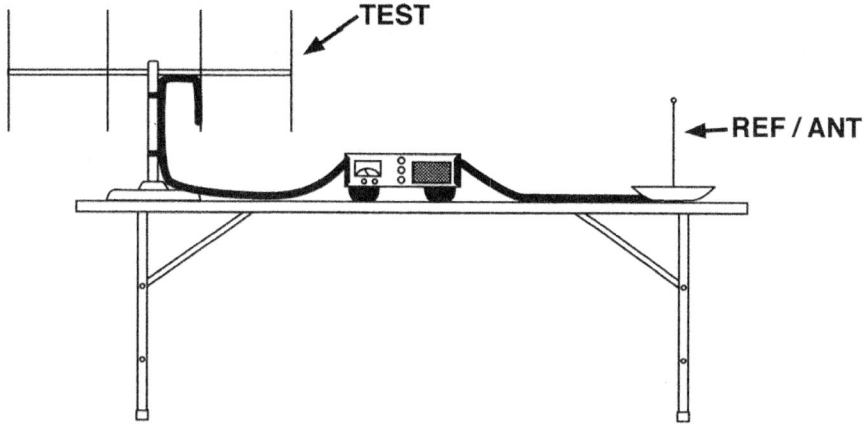

Chapter 5

Bandwidth

Chapter note: It is assumed you are familiar with an HF radio.

A resonant antenna, such as a half-wave dipole, is generally adjusted in length tuned to a specific frequency. It therefore has a certain input (feed point) resistance at this tuned frequency.

Deviating from this center frequency, either up or down the band as you rotate the main tuning dial on your HF transceiver, will displace this resistance. In doing so, the turning of the dial causes resistance changes and reactance is introduced. Both of these changes affect the input impedance. The ham radio frequencies up and down the band from the center frequency, that reach near 2:0.1 SWR at either end of the center frequency, is called the serviceable bandwidth for any antenna. It is within this bandwidth you may safely operate a 2-way radio without fear of burning up the finals in the transmitter. Thicker wire will have somewhat of a broader bandwidth than a thinner conductor will. You may operate safely at 2:0 SWR but do not operate higher than 2:0 SWR.

Center Frequency

14.150mc   14.250mc   14.350mc

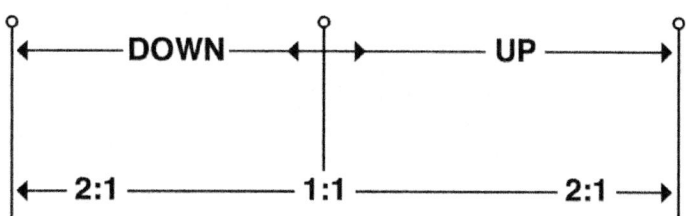

In general, antennas made of thicker solid or stranded conductors have smaller impedance variations for a given frequency displacement from center band and therefore have wider range (characteristics) than thin-conductor antennas. Therefore, with most broadband antennas, the impedance match is not greatly changed over a wide frequency band range equal to a few percent of the center frequency.

Some care must be taken when designing multiple band antenna (more than one frequency of operation), which tend to narrow the bandwidth appreciably relative to a single driven mono element.

If you are using a dipole and the band closes, operation stops until it re-opens. However, if you are one of the lucky few who has a rhombic, Texas long-gun, or other antennas that are not mentioned in detail in this book—such as a lazy-H, scatter ionospheric Yagi array, Big-L in phase, multi-dipole array in phase, and some others—you will still be working DX when others of lesser antennas cannot. These are all monster gain antennas. When this happens, some operators will hear a ham working what seems to be *dead air*, only hearing one person.

When the band closes out for most hams around the world, the few hams that are the lucky ones with the forementioned antennas above will still be working DX for minutes or even hours to come, all because of the high gain these super large antennas employ.

Further, in the following chapters the author will discuss how to build some of these monster antennas. The cost to build a monster antenna is expensive but well worth every penny.

Some hams are content with working the world on just a dipole. The author believes they do not know what they are missing. Every ham, if room permits, should have at the very least a tri-band, three-element beam for ten, fifteen, and twenty meters at least. Your radio is only as good as your antenna. It is unbelievable what you can hear. Would you buy a race car and equip it with a four-cylinder engine? A race car is only as good as its engine, and a radio is only as good as its antenna.

To give you an example, one day in the '70s the author was working MARS traffic on twenty meters from Germany. He was working with a 3-element tri-band beam. When the stateside Army MARS station began to fade out. He switched over to the station's 3-wire *resistored* Texas long-gun (wire height at twenty feet above the ground and cut for twenty meters at the thirteenth wavelength that is 866 feet/288 yards long with 1000-watt resistors at the far end). WOW! What a change. His signal on the beam was almost gone but on the long-gun, the stateside station came back up to ten over and he continued to work for about another hour before the audio completely faded out altogether. This is the difference extra *land* can make for you! A long wire is best if used in the same fashion as below. A terminating resistor may be used if desired.

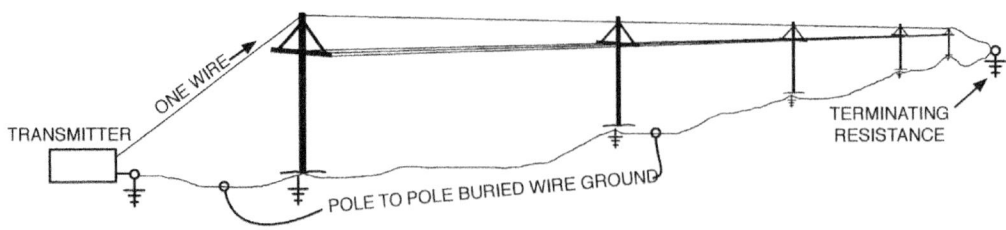

The drawing of a Texas long-gun

Chapter 6

Reciprocal Relation between Tx/Rx

The author had his doubts about writing this section. But after a long thought, he thinks the wise choice was to include this information.

An antenna may be used to transmit while a second antenna can be used to receive. However, you may elect to have only one antenna to transmit and receive on as most hams do. Let us say you want two antennas, one for transmit and the other for receive. Why would one want two antennas?

If you are prone to noise (QRM/QRN) interference, then the author recommends using two antennas. Your receiving antennas can all be connected to a coaxial rotary switch. When the band becomes noisy, simply turn the switch. One of the antennas should have less noise.

Example: Sometimes a weak signal becomes lost in the hash of noises from time to time. If you are on 80 meters listening, then switch the receive antenna over to 40-meter antenna or even a 20-meter antenna and so on; the reception, for the most part, remains the same but the noise occasionally goes away. This is an old timer's DX trick. Reducing the RF gain control may also help. Try it; you may like it. Nevertheless, it might save you from missing that one rare DX contact that you would otherwise miss out on completely.

However, if you are like most operators who have only one ham HF radio and it is a transceiver, you are stuck with the high noise. You might as well go play a round of golf or you can listen in on the 40-meter Swap & Shop on the weekends by tuning to 7.240 Mc plus or minus around 1800–2100 hours (GMT) for the International Ham Radio Swap & Shop net. If you live in America, try listening between 12:00 and 1:00 p.m. Pacific time. If you do, you might luck out and buy yourself a shortwave receiver. The author assumes that the receiver in your transceiver will mute. If your radio will not mute, then you are just out of luck all the way around. However, if it will mute then engage the mute, plug in the receiver, and try your antenna switch when noise presents itself.

An antenna may be used to transmit, receive, or both. The characteristics of radiation pattern gain, and radiation resistance apply equally well whether the antenna is used for one purpose or the other. A receiving antenna can be beamed to be responsive to an incident field over only a small angle, just as a transmitting antenna is made to radiate a field over a small angle. In this way the receiving antenna will absorb more power from a passing transmitted wave of energy than if it were not beamed—a fact that is reflected, as in a transmitting antenna, with higher directivity gain.

This information about frequency waves is true. Waves are all around us. From morning till dark, twenty-four hours a day (that's right), we are bombarded constantly by invisible rays of waves ranging from light waves, gamma ray waves, radio waves from all over the world, and many other waves too numerous to mention. At this moment, all the waves in the universe and on earth that are being transmitted by whatever means are striking you right now as you read from this page—that is *every single one of them*! If you had a receiver (with the right kind of mode) tuned into any one of these specific waves, you should be able to hear it.

All these waves have their own frequency, and they travel near the speed of light. They aren't strong enough to harm you like X-rays can, and we tend not to notice. Of all these signals, hams are mainly concerned with the ham bands, of course.

The author says, "Go out there and latch on to a receiver and build yourself a dipole from this book cut to the frequency of your choice and have fun."

Any length of antenna wire will work well for any given SWL shortwave receiver. However, due to propagation inversions, the signals may be prone to fade in intensity whereas a tuned antenna, cut to a specific lowest frequency, will capture the signal far better and reduces fadeout. This applies to receivers only.

Remember: It is plausible that a transmitted radio wave *may not* take the same path through the ionosphere in both directions of transmission due to the nature of reflections of signals as they bounce back and forth from space to earth.

A horizontal transmitted wave pattern could be reflected from the ionosphere in a vertical pattern or both. According to the US Army Antenna manual:

> "However, any observed *polarity departure reversal* from one station to another station is more likely to be caused by differences in the type of equipment used or noise level at the either of the two station locations. An invisible passing wave WILL induce a current in a tuned radiator (antenna) that is properly cut (constructed) to a half-wave (468 divided into the frequency) placed in the proper plane of polarization. As seen, there is 30 to 45% attenuation loss in signal strength between the polarizations of horizontal and that of vertical planes." *(Reference: US Army Manual dated August 23, 1956, Ft. Lee, Va., Antenna Proving Grounds.)*

When a receiver has the same input impedance as is seen across an antenna feed point, it is said to be a match circuit to the antenna's impedance or the same as saying a good SWR match.

The author hopes he has explained clearly to you at least some of the basics. Now the remaining chapters are on the construction of antennas. Enjoy!

Note: What is the "Q" factor? The Q (or bandwidth) of an antenna is the measurement of the bandwidth of an antenna relative to the center frequency of the bandwidth. Antennas with high-Q are narrow-banded, antennas, while a low-Q are wideband. A mono band dipole will have a low-Q while a trap or coil placed in an antenna will have a higher-Q.

Chapter 7

Easy up Antennas

SIMPLE DIPOLE

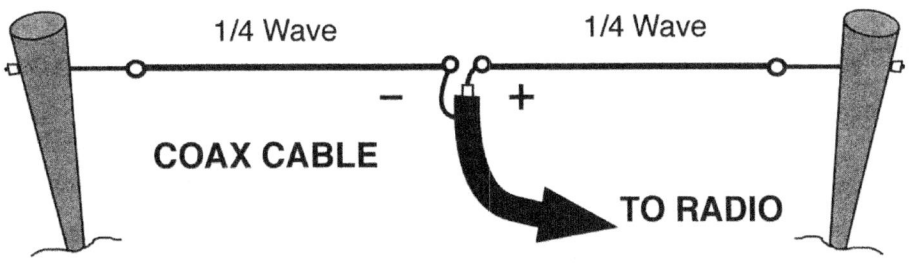

A simple dipole antenna is shown in a horizontal plane.

All antennas cut to: Antenna length 468 ÷ f = ½ wavelength

Where f = desired frequency

Sloper Dipole

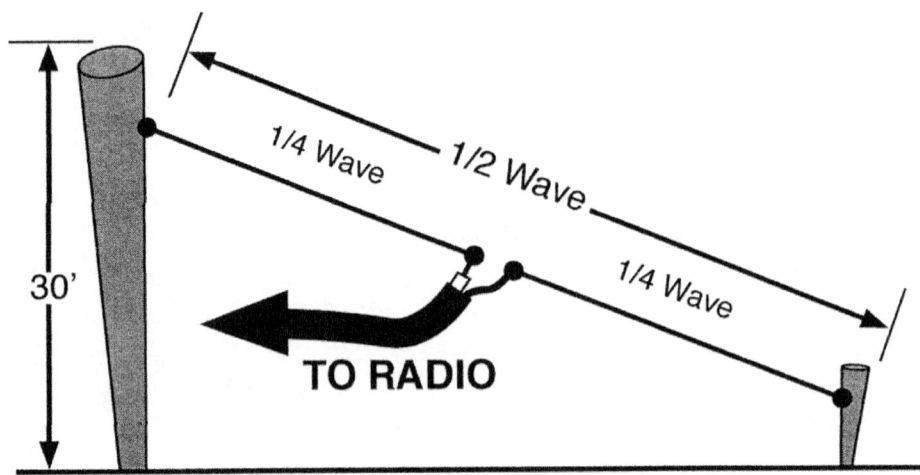

Let's say you want to construct a half-wave dipole for ten meters. The frequency you chose is 28.400 Mc.

If you already know how to do the math for building a dipole, please move on to the next chapter. For those new to the world of antennas, let's learn some more basics.

Grab a pencil and paper and let's have some fun working out the math for a dipole antenna using 468 divided by the frequency. This is the standard formula for finding the lengths for any half-wave antenna.

Get out your trusted calculator. Enter 468 then press divide. Now enter the frequency we already established being as 28.400Mc. Enter 28.400 in the calculator and hit equal.

Your calculator should read 16.478873; if it does not, then retry all your entries.

With 16.478873 showing, press the division button again followed by the number 2. (This figure 16.478873 is the total length of a ½ wave and must be converted into a ¼ wavelength.) Now press the equal button.

Your calculator should read 8.239436. This is the length of each of your 2-each, ¼ wave wires.

You will be cutting two separate lengths of wires that, when assembled, will be overall the total length of a simple dipole from end to end at a half ½ wavelength.

In order to un-reel two sections of number 12 or 14 gauge insulated electrical (copperweld) wire from a spool that you will be using for the design of an antenna, you will later on need to measure each wire to a ¼ wave section with a 50- to 100-foot measuring tape, one section at a time.

The calculator reads 8.239436. Take note where the decimal is located.

$$8.239436 = ¼ \text{ wave}$$
Left ¦ Right FEET ¦ INCHES

The number 8 is on the left side of the decimal and is the footage.

Remember: All numbers on the left side of the decimal point are in feet. All numbers on the right side of the decimal point are in inches.

On a piece of paper, write down 8 feet. We now know that the given length of our ¼ wave section for the ten-meter frequency of 28.400Mc is 8 feet, for a receive antenna. For transmit, we must continue. Now we must figure out the inches and then figure out on down to the 16th of an inch.

For a transmit antenna, the antenna must be cut precisely. As for a receive antenna, it does not.

With the calculator showing 8.239436, press the minus key, then press the number 8 followed by the equal key.

The calculator now reads 0.239436.

Since we know that the number 8 is the footage, and we saved it by writing the number 8 down, we now delete footage from the calculator and move on to find the inches.

Enter the number 8 followed by a decimal point then press the equal key. The 8 will go away. What you have showing now, on the calculator, are the inches—but these inches are in the form of decimals and must be converted into inches. We will now convert .239436 from decimals into inches.

With .239436 showing on the calculator, press the multiply key, then enter 12. Now press the equal key (there are 12 because there are 12 inches in one foot).

Now look where the location of the decimal point is.

:
2. 87328
LEFT: RIGHT
INCHES: FRACTION of an inch

On the same piece of paper, write down the number 2 next to the number 8 previously written. Now delete the number 2 from the calculator by pressing the 2 keys, then press the decimal key, and the equal key.

You now have 8 feet and 2 inches, known. Now we go after the 16th (sixteenth) the same way as we did before with the others.

The calculator now shows 0.873238. Some people stop here while others continue and multiply the 0.873238 by 16. Your choice.

You now have the working length to build up a dipole antenna.

Remember: A dipole has a zero gain in real life. However, when compared to an isotropic radiator, the dipole presents a gain of +1.5dB. (Read Author's Thoughts on Gain.)

A dipole antenna cut to a half-wavelength and mounted above the ground at or near 65 percent of a wavelength in either a vertical or horizontal plane presents near 73 ohms at the feed-Point.

To display proper current distribution uniformly, use a 1:1 balun.

UNDERSTANDING HF WIRE ANTENNAS

Remember: Current and voltage distribution of a dipole is as follows: Current is at its maximum at the center-feed point and minimum current at the far ends on either side of the feed point. Voltage, on the other hand, is at maximum at the far ends on either side of a dipole and minimum at the center feed point. A dipole is, for the most part, independent of the ground once it is suspended 65 percent of a wavelength above the ground.

Folded (Dipole) Doublet

A folded dipole is essentially the same as a half-wave horizontal, except that its input impedance has been increased to permit feeding it through a suitable open-wire transmission line or coax cable with a 6:1 balun. The total length of wire in a folded dipole is in multiples of half-waves.

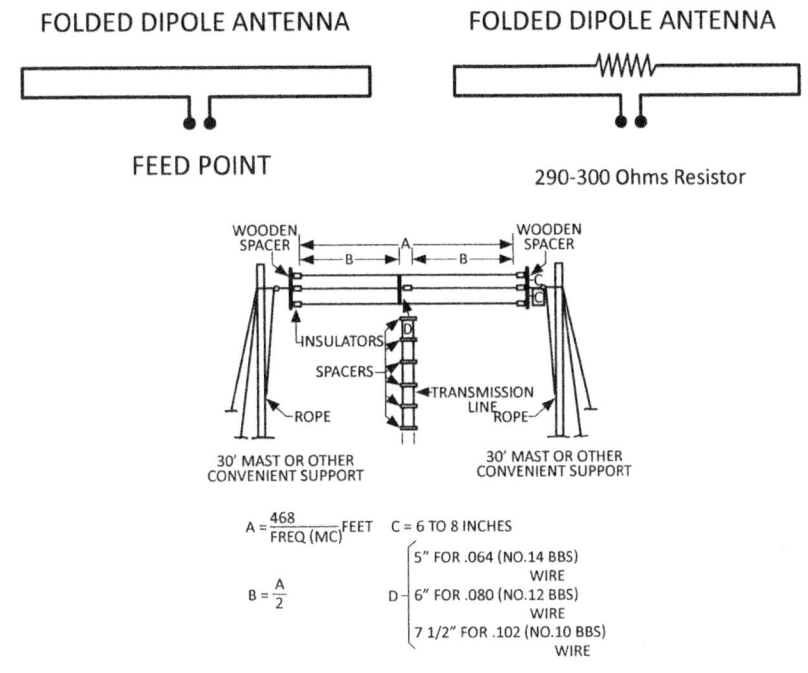

Three-wire folded-doubled antenna

Inverted Vee

The difference between an inverted dipole and the inverted vee is simply an inverted dipole is a half-wave in nature and the inverted vee consists of many quarter-waves cut to odd multiples.

When you invert a dipole, inverted vee, or any other antenna at the center feed point, it is said that the center feed point now becomes known as an apex. When you hear or read of an apex, it is referring to an angle-point of an antenna and normally associated with a feed point. The two wires arranged in the form of a V when the apex angle is optimum, provides good directivity and gain characteristics for a single band operation. For multiband, add termination resistors at the ends on each side and then use an antenna tuner.

Feed the apex of the vee by means of a balanced feedline such as ladder-line or unbalanced using a balun and coax cable. The apex angle is so chosen that the main lobes of the wires reinforce each other along the bi-sector of the apex angle. The lobes in other directions tend to cancel out.

Design inverted vee by using 468 ÷ f is okay. When no resistors are used in the inverted vee, it becomes bi-directional. The main lobes will be display from both sides as in the drawing below and the overall gain is split between both sides. Using resistors turns the main lobe from both directions into one power lobe, like that of a beam.

Place resistors at the far ends of each side of the wires.

UNDERSTANDING HF WIRE ANTENNAS

Resulting Patterns of an Inverted Vee Directivity of Bi-Directional/Uni-Directional

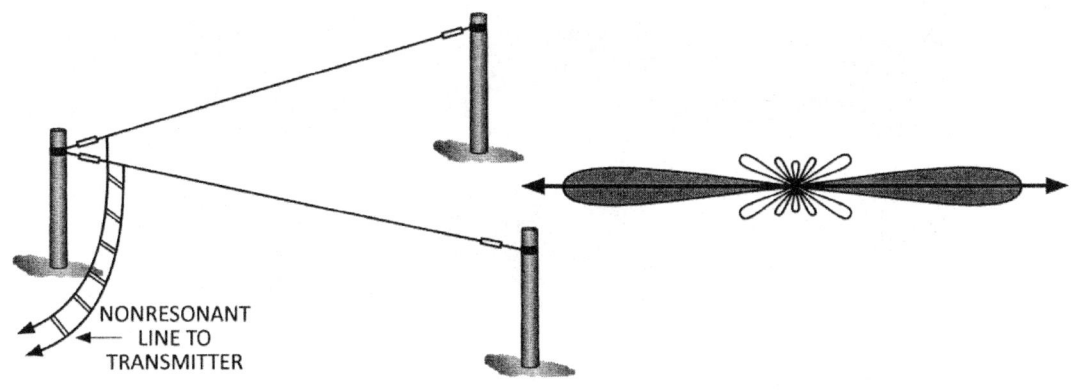

A standard vee-type antenna is a resonant bi-directional antenna.

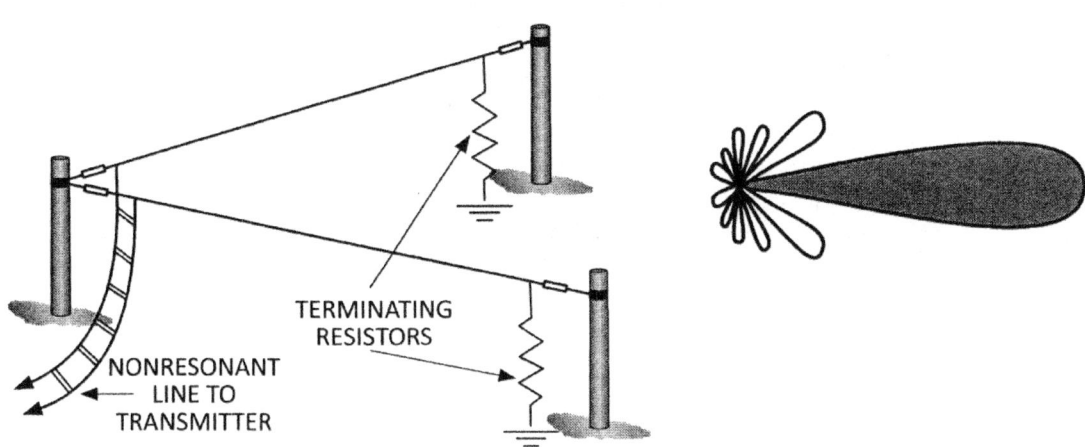

The placement of terminating resistors on each legend will turn the inverted vee into a uni-directional non-resonant antenna.

Wave Beaming Power

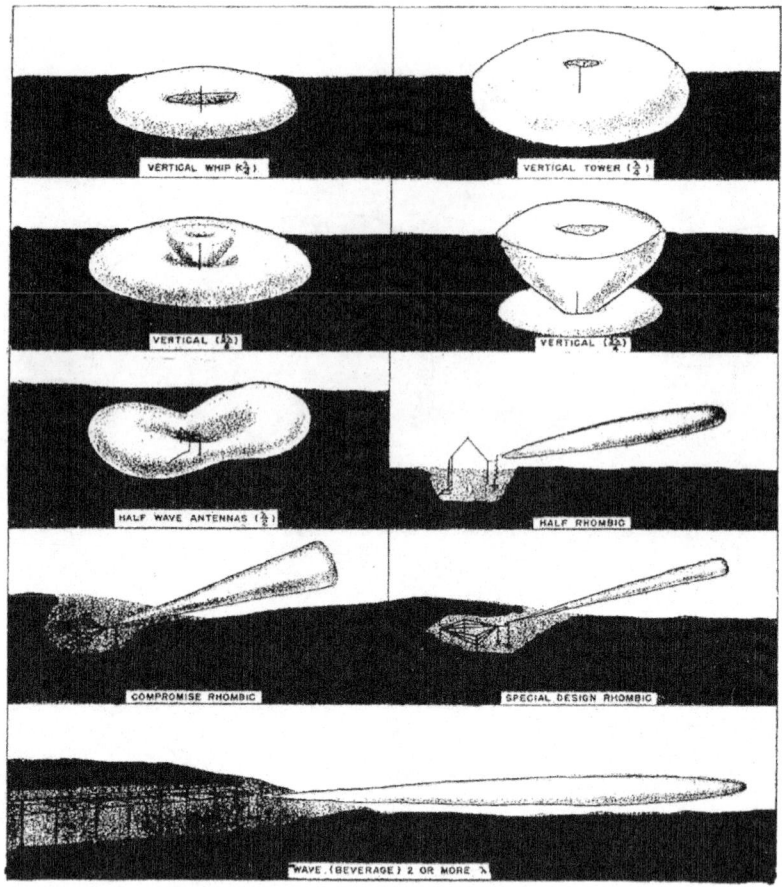

The narrower the beam, the better the gain

Chapter 8

Basic Monster Antennas

Long Wire and Texas Long-gun

Long wire and long-gun antennas are mostly designed for *single band operations* and should never be used out of band with any antenna tuner. However, multiband usage with an antenna tuner may be operated safely if *termination resistor(s)* are added to the wire at the far end to ground.

Long wire and long-guns are known as a wave or beverage antenna. Long-wires are exactly what they are—a single long wire suspended above the ground ten to thirty feet, energized with a current and voltage by a transmitter to produce a wave pattern that will function as an antenna. It can be designed to be either a resonant or non-resonant. Resonant wires are cut for a specific single frequency depending on the desired wave pattern as seen in the drawing below. Cut all long wire antennas in half-wave multiples.

The pattern tends to shift its maximum lobes more and becomes narrower toward the end of the antenna as the length increases. The length works best between ten to twenty feet. You can go higher, but do not go higher than thirty-five feet.

Resonant antennas are designed for a certain frequency in mind whether it is a dipole or up to and including a long-gun to rhombics and should never be used out of band with any antenna tuner. In doing so, excessive unregulated current may run wild through your tuner and into your antenna system causing severe damage to your shack, antenna, and property. Runaway voltage and current will produce heat within the antenna network, which includes a transmitter, turner, and feedline and dissipate at the weakest point.

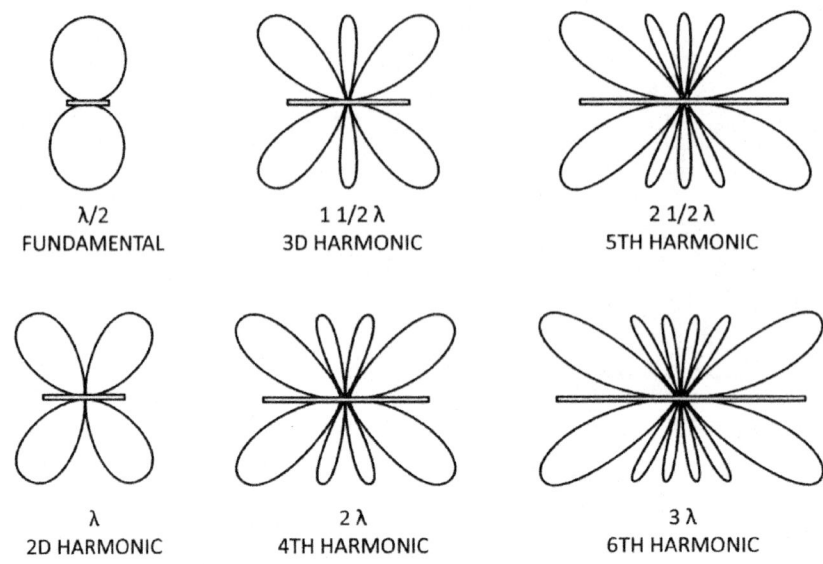

Wave pattern shown non-resistored – Resistored brings all lobes beaming in one direction

Remember: An inverted vee is also a balanced antenna; that is to say, both wire sides are of the same length—but it is not a dipole! A dipole is designed to a specific frequency and can also become an inverted dipole by sloping the wire ends toward the ground but not touching the ground. Whereas an inverted vee are that both sides of the antenna wire are of the same random lengths measuring to the lowest frequency you wish to operate and requires an antenna tuner to load.

UNDERSTANDING HF WIRE ANTENNAS

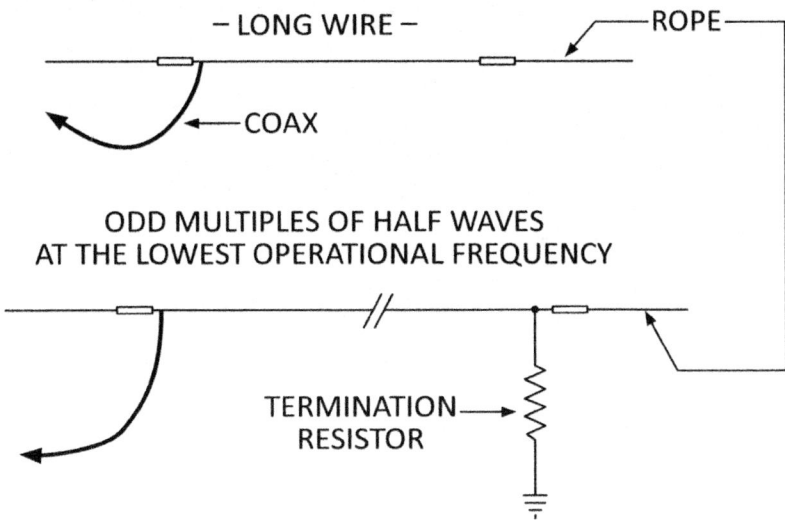

A signal is attenuated when using an antenna tuner on a resonant antenna out of band. A long wire can be designed so that there is no reflection from the ends of the antenna wire; this prevents melting of wires at the ends. It also prevents melting the supporting insulators. Adding a termination resistor guards against unwanted extreme amount of dissipated heat, induced by massive unregulated current.

When termination resistors are introduced in the antenna circuit, it becomes a non-resonant antenna. Reflection is therefore eliminated by arranging the conductors so that the far end of the long wire is terminated by adding a terminating resistor to the antenna circuit. See Termination Resistors in Chapter 10.

These terminating resistors must be equal to but not over .05 percent of the impedance of any wire antenna system. When operational, the antenna then is operable over a wide frequency range with an antenna tuner.

Determining Long Wire Length

The length, like that of a half-wave antenna, is modified by the end effect of the antenna wires. This end effect is formulated to determine the length in feet of a resonated long wire antenna for any given frequency in (MHz, mc) megacycles:

$$492 \times (H - .05) \, L = F$$

Where

L = length in feet

F = mc

H = half-wave

.05 = Percent of a wavelength

It now becomes apparent that the antenna will be slightly off resonance at various harmonics of the frequency for which it is cut. The discrepancy is negligible at normal harmonic frequencies in the HF band.

Feeding the long wire is not critical due to the nature of their design.

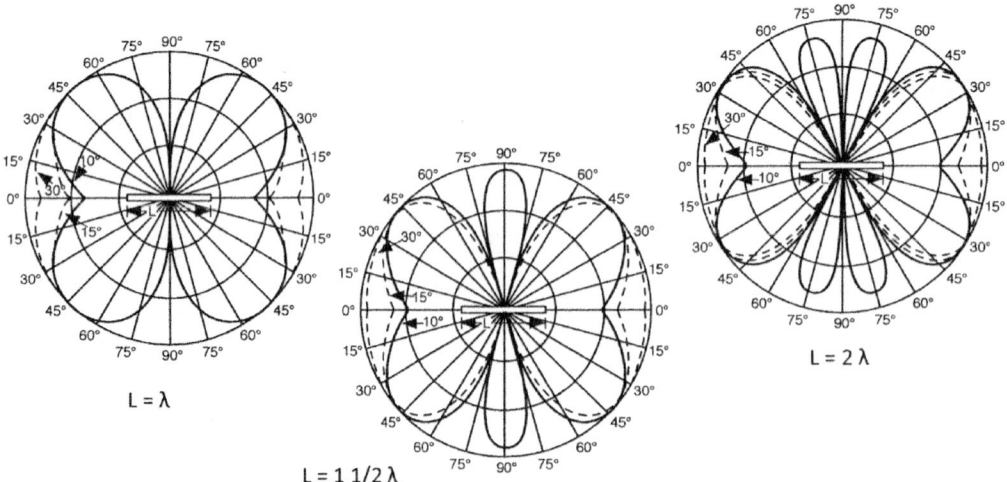

Long wire radiated wave pattern in wavelengths

Undisputed polarization pattern of a long wire is a combination of both the vertical and horizontal planes. Most hams realize this and use it as a backup for DX listening as their primary or even as a receiving antenna.

The radiation pattern and practical gain of a long wire radiator over typical soil differs from the free-space radiation pattern and gain. They are usually different from the pattern and gain that would be obtained with the same radiator over hypothetically perfect ground conditions.

The azimuth (horizontal plane) directivity will vary widely with the vertical angle involved. The pattern obtained with the antenna over typical ground is further complicated by the fact that directly off the ends, the radiation is vertically polarized; however it is horizontally polarized from the broadside of the antenna, and at intermediate angles, it has both vertical and horizontal components. In short, a long wire will capture all patterns of waves at the same time unless the long wire employs a resistor.

Connection: Texas Long Guns and Long Wires

Texas long-guns and long-wires; at the feed point, all antenna wires are connected to the center wire of the feed line coax or one side of the ladder line. The negative braid of the coax or the other side of the ladder line connects to the pole-ground-network. A long-gun antenna may be used with or without a balun. Both of these antennas require an antenna tuner capable up to 90 to 2000 ohms or more. It is better to be over than not enough. Connect coax center wire to antenna wire and coax ground braid to grounding network at the feed point.

If using ladder line, feed it just like a long wire setup. At the far end, all antenna wires are connected directly to one side of a termination resistor and the other side of the resistor is connected to the pole-ground-network. This will allow you to operate all bands down to the lowest frequency of the antenna.

If no *terminating resistor* is used at the far end of the antenna, the antenna wires are then connected to each other and will not be connected to anything else; this is not the case for rhombics. This setup is for single band operation only.

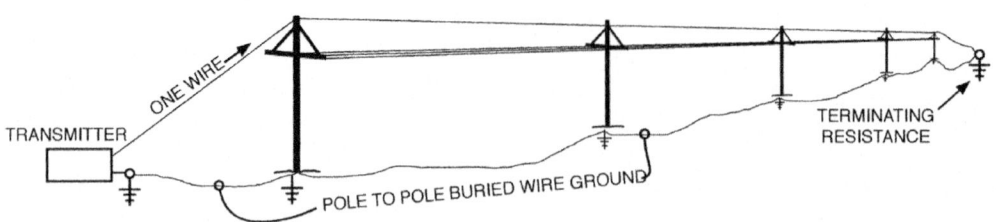

End Pole of Long-Gun

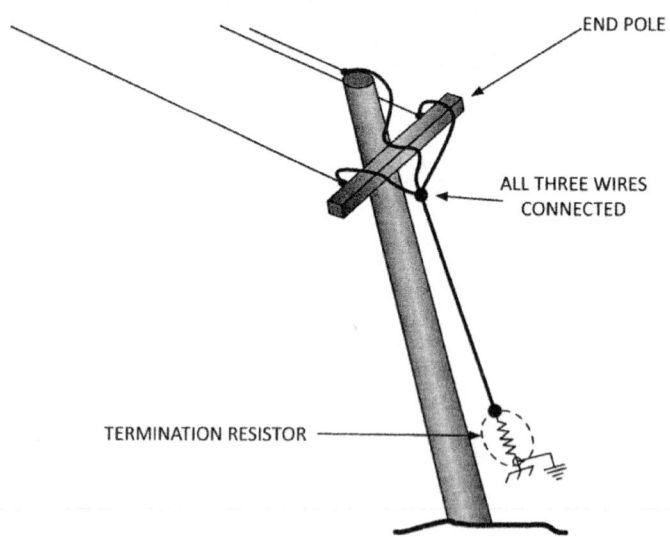

This is the end pole of a Texas long-gun. If you terminate the end with a resistor, you will have a non-resonant antenna. Using a minimum of 135 feet or more will allow you to operate on any band with an antenna tuner capable of handling 90 to 600 ohms.

Without a resistor, just short all two or three wires together and leave it alone, that is, not grounded. This setup is a resonant antenna for a single band operation only. Many hams build antennas without being resistored and tune-up out of band causing damage to their antenna tuner. Some will flat out smoke it while others might just cause pitting from current arcs.

Long-Gun Mid-Support Pole(s)

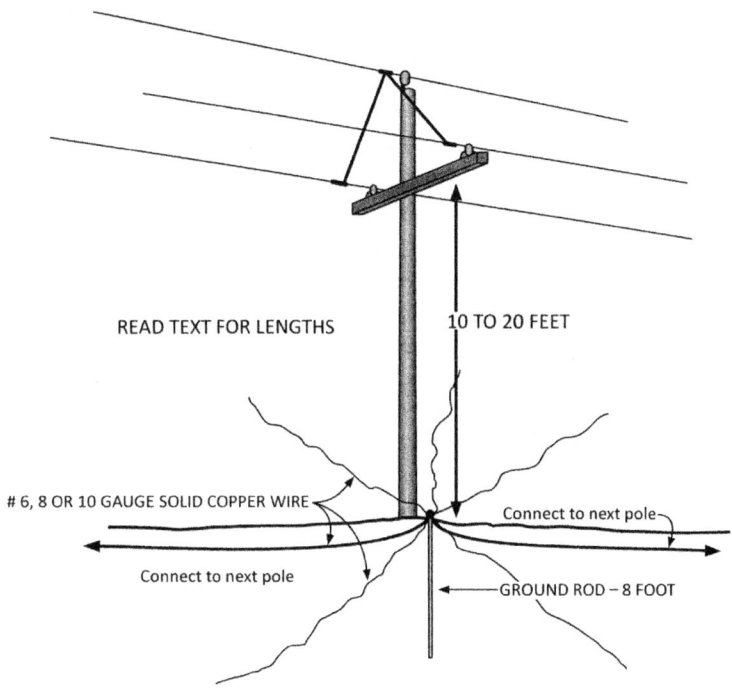

Buried Grounding Network for Long-gun

All ground wires are shorted together at each pole and grounded.

Diamonds and Rhombics

A rhombic, Texas long-gun, and a diamond are monster antennas that can have awesome gain especially if you have two or more co-phased. The gain of any antenna is normally compared to that of a quarter-wave or dipole antenna, with the same polarization at the same height with one watt at one mile.

A single-wire (2 wire or 3 wire) stranded rhombic antenna designed for 13 wavelength presents a power gain nearing that of 8 to 10dB true gain or more over a dipole or 11.5dB over isotropic radiator depending on length, height, angle, and usually terminated with terminating resistors at the far end causing the antenna to become uni-directional. The main beaming power will be projected in one direction only. If you are not using terminating resistors, the beaming power will be half the gain mentioned in either direction. The main beaming power will not be in one direction but in fact in two directions, both forward and backward. Therefore, dividing the total gain in half (bi-directional).

Adding more wire within your present total length does not add more gain but does decrease the overall impedance, such as 1-wire, 2-wire, or 3-wire antenna systems. When adding more wire, at the feed point, 2-wire and 3-wire are connected on each leg side and spaced on the center support pole by a foot for each wire. At the far end, all wires are again connected together for each side legs. (See rhombic drawings.)

Adding more wire to extend the total wavelengths will add more gain and this makes the antenna longer. The gain is insensitive to the conductivity of the ground but is associated with a given vertical angle in a horizontal plane. A rhombic is normally designed for receive and when employed for transmitting multi-frequencies must be designed for the lowest frequency that will be used. An antenna tuner must be used.

Remember: Terminating resistors will allow multiband use. Without it, you have single band use. These monster antennas could or will require a tuner that can operate within the ranges of 500 to 3000 OHMS! Consult your antenna tuner manufacturer.

UNDERSTANDING HF WIRE ANTENNAS

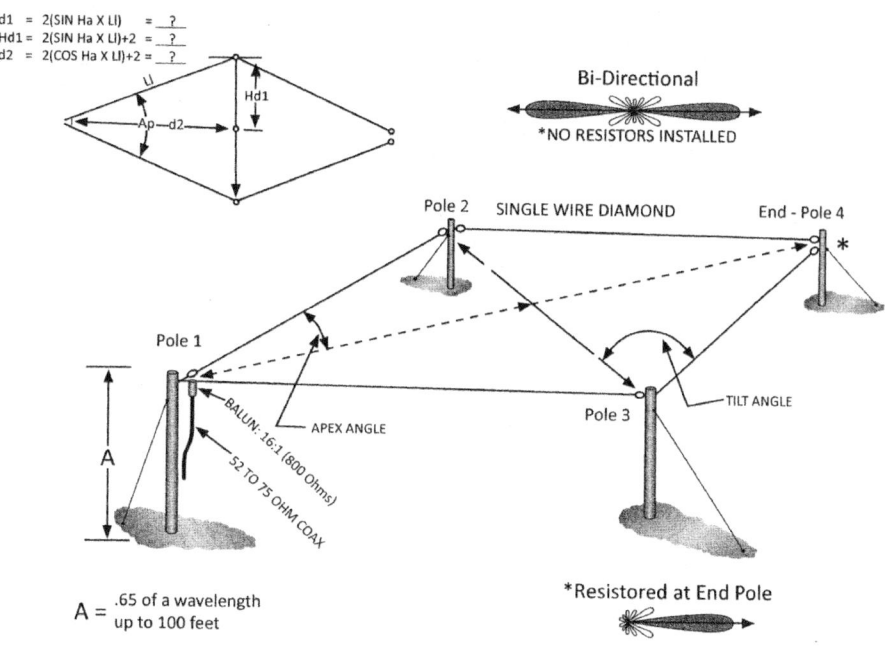

Single-wire diamond

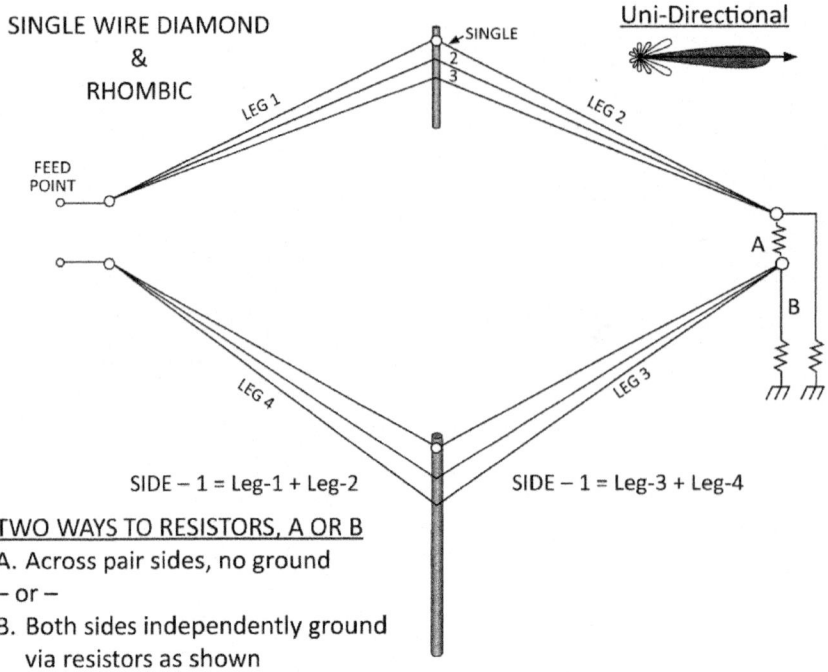

Rhombic Characteristics

Characteristic impedance of the rhombic antenna is somewhat constant over its entire length. Input impedances will vary little, but close to 800 ohms for a single-wire diamond rhombic antenna, 725 ohms for a 2-wire, and 650 ohms for a 3-wire rhombic over a frequency range of 4 to 30 Mc. Be sure to test your impedance with an antenna analyzer to be certain.

Leg length, usually expressed in wavelengths, is the length of each one side of the four wire legs that make up the four sides of a rhombic.

The tilt angle is one-half the inside angle between the wires at the side poles of the antenna.

Range in miles	Side length (feet)	Tilt angle ° (deg)	Height (feet)	Length, end pole to end pole (ft)	Width, side pole to side pole (ft)
over 10,000	425	90	90-100	825	310
over 3,000	375	70	65	723	268
2,000 to 3,000	350	70	60	676	251
1,500 to 2,000	315	70	57	611	228
1,000 to 1,500	290	67.5	55	553	234
600 to 1,000	270	65	53	506	240
400 to 600	245	62.5	51	453	238

Height is expressed in wavelengths or fractions thereof. This is the distance between the plane (wire) of the antenna and the plane of the average ground level.

Angle of maximum radiation: This is called the wave angle. The wave angle should coincide with the angle of radiation or arrival of the optimum propagational path for the antenna's electrical circuit. The wave angle can be used to determine the three dimensions of the rhombic antenna and the tilt angle. For any one wave angle, one set of these dimensions gives the maximum output at the particular angle.

Horizontal/Vertical Patterns: Directivity patterns for any typical rhombic antennas are shown in the drawing below. They apply to either a transmitting or receiving rhombic antenna. The transmitted or received signals are amplified within the directed main lobe. The principal lobe is in the general forward and upward direction; that is to say, starting from the feed point at the apex then out toward the opposite in or at the termination resistor, should that be the case.

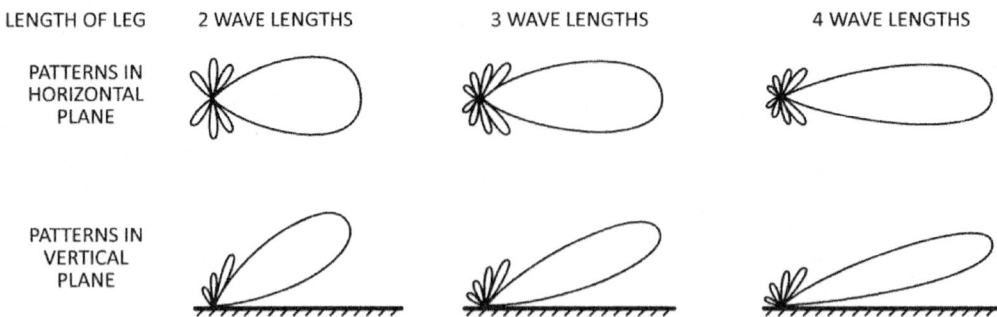

Elevation: One Wavelength

The radiation pattern is dependent upon the length of each side of the antenna, the tilt angle, the frequency, and the height above the ground. Since the angle required between sides do not vary appreciable with frequency changes, and since the circuit is non-resonant, a wide range of operating frequencies may be used.

Connection

These large antennas are referred to as wire gain antennas. More than one wire setup is called *curtains*. Adding more wire *length* will increase gain. If the far end of the antenna is left open—that is to say, not terminated by means of resistance and not shorted to each other side—the antenna power lobes will be bi-directional.

The rhombic is now a resonant antenna that is tuned to a specific single frequency. Therefore, it should not be used out of band with a tuner. High voltage and current dangers will be present at the antenna tuner when attempting to tune out of band.

If the far end of the same antenna is terminated with termination resistor(s), a high impedance antenna tuner, capable of handling up to 2000-plus ohms, may be used without (much) problems. always the antenna tuner grounded! Using resistors, the rhombic will display a uni-directional (monster) beaming pattern in one direction. This statement applies to all antennas using non-inductive resistors.

There are two ways to add resistors:

- Take sides 1 and 2 at the far end and connect it to one side of a single resistor. The other side of the resistor is connected to the other side of leg 3 and 4. In this setup, there is no ground connection.
- The alternative connection uses two resistors each. Take side legs 1 and 2 and connect them to resistor 1. The other side of the resistor 1 connects to ground. Repeat this process with side legs 3 and 4 to resistor 2 to ground.

Chapter 9

Construction of Antennas

Safety is *paramount*! Whatever you do or however you do it, proceed with safety in mind, and always stay away from nearby power line wires!

Most antenna books omit the following construction techniques. The author included these techniques for the novice ham operator. When constructing larger wire antenna systems, failure to understand the following ten items could be very costly to your pocketbook. These steps are especially important!

1. How far can I space my antenna support (power / telephone) poles? How deep must each pole be planted in the ground in order to be safe?

Mount antenna wire on support poles at ten to thirty feet above the ground for long-wires and long-guns spaced in intervals of 50 / 75 / 100 / 150 / 200 / 250 feet apart accordingly. Place antenna wire for diamonds and rhombics at forty-five to one hundred feet above ground and spaced by reading charts and drawings.

The wire for theses antennas will be placed in a level plane, line of sight, no matter what the terrain is below the wire. The number of poles should be selected as required for the job according to the total length of your project. Pole identity should be as follows: feeding pole, turn poles, and end pole.

For long-guns—one 4x4 cross-arm, three double-through arming long bolts with three support braces, eight square washers, and one tie-down guy-wire kit for all poles to be used (consult your power company).

For diamonds and rhombics, use two double-through arming bolts with bracket attachment for antenna wire tie-downs at the feed point. All others use one double-through arming bolt for each wire attachment. For a diamond, use a 2-wire rhombic, and use two bolts. For 3-wire, use three bolts for the support middle poles. At the end pole, use two bolts for wire tie-up. Use one bolt for the leg side 2 and 3, and the other bolt for leg side 3 and 4. Use guy-wire tie-down hardware for each pole if needed.

Considering Pole Spacing Minimum to Maximum

Support spacing of poles becomes a factor in situations where conductors (antenna wire) may sway or swing in the wind or coated with ice, causing costly damage. Mechanical strength rather than current-carrying capacity is a factor such as with gold-plated copper wire rated at number one, silver next, then copper, and inferior metals last. Cost is a deciding factor in any final determination of what type of wire is employed. But cheap can be costly. Think about it!

2. Wires having a low tensile strength *cannot* be stretched tightly enough without breaking while using a come-a-long tool for proper strand (wire) tension between your support poles (should that be the situation). Breakage due to high winds or under heavy weight from ice may occur. Wire that is too long between support poles tends to sag, thus becoming susceptible to sway, causing insulators to snap off from supporting poles at the cross-arm in high winds. When this happens, you will have to hook the pole and replace them. All large high gain antennas such as the Super long wire, Texas long-gun, 1-2-3-wire

(curtain) rhombic, or any other larger type antenna will use for the most part 1/8, 1/4, or 3/16-inch steel strand.

3. If these heavier cables as mentioned above are not in your construction plans, then use gauges such as number 10 and upward. Use copper-weld wires; these wires are designed to keep the stretching at a minimum. Just keep in mind that whatever wire gauge you choose to use from 14 gauge up to 3/16 inch for your antenna system; the wire must not be able to stretch. If it does stretch, it will affect your SWR.

When placing the wires along the support poles during the construction, make certain that you closely observe that there are no kinks (where the wire becomes knotted or twisted). Avoid kinks at all cost! Current, when introduced into the antenna, could possibly stop at the first kink, causing high SWR. This could cause damage to your antenna, antenna tuner, or even to your transceiver. Each kink will react as if it were an in-line resistor and might stop the current altogether or limit your voltage and current at that point. This defeats your whole purpose.

Avoid sharp turns at the corners of the antenna. This too will create resistance and hamper the voltage and current flowing on the outer side (skin effect) of the conductor (antenna wire).

Make sure at the corners, such as with a rhombic or diamond loop, the angle of the wire at the turn are in a smooth curve coming from one side and going out toward the other side. Example:

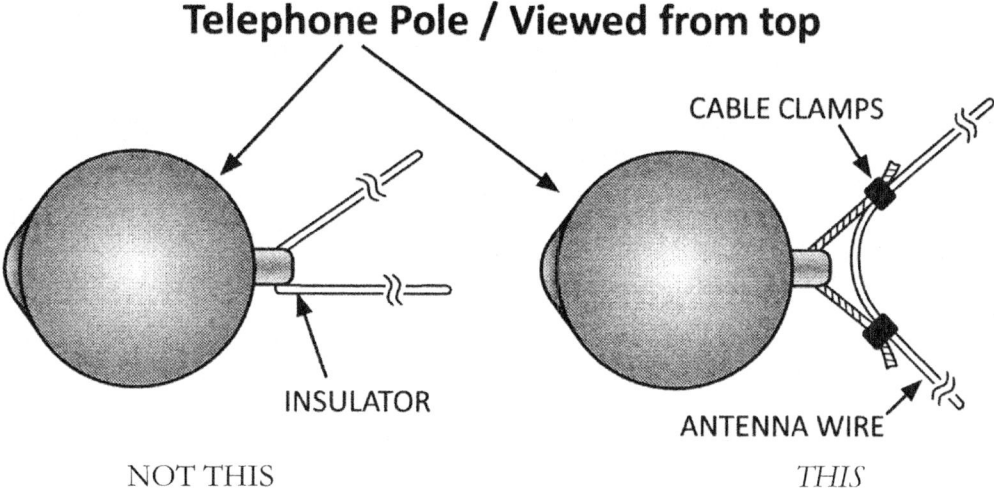

Telephone Pole / Viewed from top

NOT THIS — INSULATOR

THIS — CABLE CLAMPS, ANTENNA WIRE

4. In the case of inverted vee, rhombic, or any angled antenna, the lengths of all side legs (legs = wire of the antenna) must be kept identical in order to exhibit a proper SWR match.

5. In a single (mono) band antenna system, direct the feedline (coax/ladder line) in such a manner whereas uniformity is the issue. For mono band antennas, cut the feedline in whole multiples of half-wave lengths using the K-factor. Any length is fine for multiband antennas.

6. All wire supporting poles should be placed at varying distances from each other. *No two adjacent pole spans should be alike.* Each insulator will add to the conductor a definite mass of conductive material in the form of tie-down hardware/wire and metal ends of the insulators. The variation non-uniform pole spans prevent the establishment of unwanted resonant sections along the antenna line by various reflections from the ground and from lumps of conductive material.

7. Attention must be given to building entrance arrangements. Coax and ladder line into your ham shack should be neat.

8. Sagging antenna wires have been a problem for many hams in the past. Sagging wires will continue to be a problem as they deteriorate in weather as the days and years pass on. All antennas require repairs over time.

Sagging antenna wire is an issue that must be reckoned with or else damage may occur during the operational life span of such systems. Anyone can put up a simple antenna and it might work well—and then again, maybe not. All hams that reach out to communicate further have had quality as their priority. *Sagging is the maximum departure of a wire below the plane of the straight line of sight between two supporting poles.* The most common tendency is to pull small conductors too tightly or to leave excessive sag in large conductors.

When a small conductor wire is pulled in too tight, the wire has a high chance or a tendency to snap during construction or in seasonal weather mainly to contraction or expansion.

Insulators have been snapped from poles due to cold weather wire-contraction, or where a lead wire (antenna wire between spans) had snapped.

9. Another consideration is a conductor wire that might stretch after years of being weathered or damaged during the construction stages from being over tightened by a come-a-long tool.

Leaving too much wire between spans will result in the wind whipping the wire back and forth. This could cause the wire to break or snap off an insulator. The following chart will show proper sag in the middle between two poles or supports. Please study it. Sag lengths are in inches. The sag will produce a tension in pounds, which is felt at each end of the supporting insulators. The table does take into consideration high ice regions as well as extreme desert regions.

This chart is good for all wire antenna spans between support poles.

SPAN (IN FEET)	SAG (IN INCHES)			TENSION (IN POUNDS)		
	30 deg F	60 deg F	100 deg F	30 deg F	60 deg F	100 deg F
100	2.3	2.8	3.9	362	288	216
125	3.6	4.5	6	360	288	218
150	5.1	6.4	8.5	358	288	221
175	7.1	8.7	11.5	357	288	223
200	9.3	11.4	14.7	355	288	226

SAG chart is averaged and good for No. 10 AWG gauge to 3/16-inch strand.

The recommended sag lengths for conductors used in heavy ice-loading areas of the Earth's temperate zones may be excessive in the tropics and in other warmer temperate climates. In polar regions with extremely heavy ice-loading and high winds, the amounts of sag shown in the chart may be increased slightly. Sag for intermediate span lengths or temperatures can be determined by interpretation.

10. All support poles including telephone, cable TV, power, and 10/20-foot support poles *must be buried in the ground at a proper depth to prevent injury.*

A pole setting guide chart is presented on the next page. Many hams have wondered, what is a safe depth for a pole carrying weight without guy line support?

This chart must be followed. Study it carefully. US military, cable TV, power, and telephone companies all use a similar chart.

Concerns should be made to maintain a constant horizontal antenna plane. Antenna wires with respect to the earth are affected by varying elevations. The pole heights shown in the chart below are seen from level ground.

Remember: Pole heights, pole cross-arms heights, and harness attachments heights must all be calculated for each pole of the proposed antenna. You must consider the ground elevation at the base of each pole, so you do not hinder the plane (that is, the line of sight of the antenna wire plane). This will aid in maintaining an overall level horizontal plane of the finished antenna, should you have uneven ground beneath the proposed antenna site.

Study the pole depth chart below carefully!

\multicolumn{4}{c}{Support poles to be used between spans}
Total length of pole (ft)	Height above ground (ft)	Depth set sunk (ft)	Approximate butt diameter (in)
16	12	4	8 to 10
20	15.5	4.5	8 to 10
25	19.5	5.5	8 to 10
30	24	6	10 to 12
35	29	6	10 to 12
40	33.5	6.5	10 to 12
45	38.2	6.8	10 to 12
50	43	7	12 to 18
55	47.5	7.5	12 to 18
60	52	8	12 to 18
65	56.5	8.5	12 to 18
70	61	9	12 to 18
75	65.5	9.5	18 to 28
80	70	10	18 to 28
85	74.5	10.5	18 to 30
90	79	11	18 to 35
95	88.5	11.5	18 to 45

WARNING! In no case should the *depth settings* in the chart be reduced for *any reason* on a stand-alone non-supported pole!

The setting depths given are at the *absolute minimum depth!* Guying the pole, on the other hand, is a different matter and the hole in the ground may be reduced but only by 5 percent—if it is hard-packed dirt!

Radial Ground System

The poorer the soil conductivity below monster antennas, the slower the wave velocity will be along the antenna for long-guns and rhombics; therefore, a compensation should be used in the form of Radial Ground Reflection Network System.

Install ground wires at every pole and link all the poles together from feed point to the end poles. Bury 6- to 10-gauge copper grounding wire between all poles. Ground the feedline—negative side or coax braid—to this grounding system. Place an eight-foot ground rod at each pole and tie it to the grounding wire. DO NOT SHORT ANTENNA WIRE OUT TO THE GROUND SYSTEM.

Chapter 10

Termination Resistor

Terminal resistors may be utilized on dipoles, inverted-vees, rhombics, diamonds, and beverage/wave antennas like that of Texas long-guns and long-wires.

Tubular Ceramic Resistors

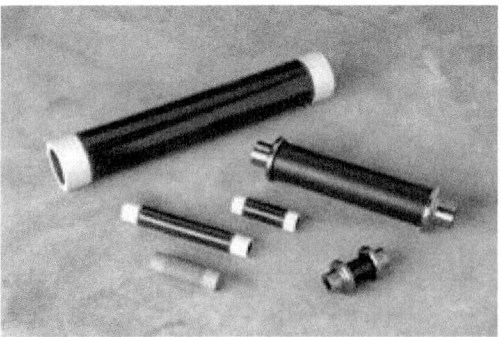

High voltage resistors – High-power resistors – High-energy resistors

Series 800 and 1000 tubular resistors are non-inductive and available in a wide variety of sizes and terminations from 2 to 24 inches in length and ½ to 2 inches in diameter. These resistors can handle up to 1000 watts, 165 KJ and 165 KV in resistance values from 1 ohm to 1 megohm. *(Source: http://ppmpower.co.uk/products/resistors/ceramic-resistors//)*

A mounting protection cabinet should be weatherproof to prevent electrical shorting from the elements. The total resistance of the termination resistors should be higher than any normal antenna impedance by at least ten percent. Use pure mineral oil to dissipate more heat if needed. The resistance may be variable in steps, so that it can easily be adjusted at ground level at the cabinet by a switching means in order to adjust the match of the antenna impedance. A remote electrical relay system may also be used.

The resistors must equal the characteristic impedance of the antenna. This allows for good grounding and tuning.

Inverted vee: When designed in odd multiples of quarter-waves over one wave in length, the terminating resistors should have a valve of approximately 500 ohms for each single leg for an inverted vee and must be capable of dissipating at least one-third to one-half the power in watts supplied to the antenna. Terminate each side leg with a resistor to ground.

Long wire resistors have funny values start out with 90 ohms working up to 300 ohms and are capable of handling the heat of one-half of the watts supplied to the antenna. Try different varieties until you find the value you favor. Terminate the far end with a resistor to ground.

Antenna length (wave lengths)	Radiation resistance (ohms)
1	90
1 1/2	100
2	110
2 1/2	117
3	122
4	130
5	138
6	144
8	154
10	162
13	170

Beverage/Wave Antennas (Like that of a Texas long-gun)

If termination resistors are used, start with 200 ohms and if necessary, work up to 450 ohms. This may take a while to find the right value.

Make sure resistors are capable of dissipating at least one-half of the wattage supplied to the antenna. At the very end of your wire antenna, terminate all wire legs to one end of the resistor and the other side goes to ground.

Rhombic antennas: A regular bi-directional—resonant—non-terminating resistor installed. A rhombic will work simply fine for transmit and receive; however, if you want a monster with uni-directional characteristics then the rhombic antenna must have a *termination resistor*. Set the resistor to the proper ohm value. The proper values of resistance are important for a rhombic—it is approximately 800 ohms for a single-wire diamond, 725 ohms for a 2-wire rhombic, and 650 ohms for a 3-wire rhombic. Build these resistor networks to handle one-half to three-quarters the transmitting power for the antenna. Terminate each end legs with its own resistor to ground (I favor this way) or as some say, you can terminate the end legs to each side of the resistor and not grounded. This method allows the top height of the lobe to rise higher than the grounded method.

Try both methods to discover what's best for you.

This setup is used for all monster antenna setups. The transfer point is always located at the end of the antenna. A rhombic or diamond will have two wires coming from the antenna to the *transfer-point* and a long-gun or long wire will have only one.

Termination Transfer Point

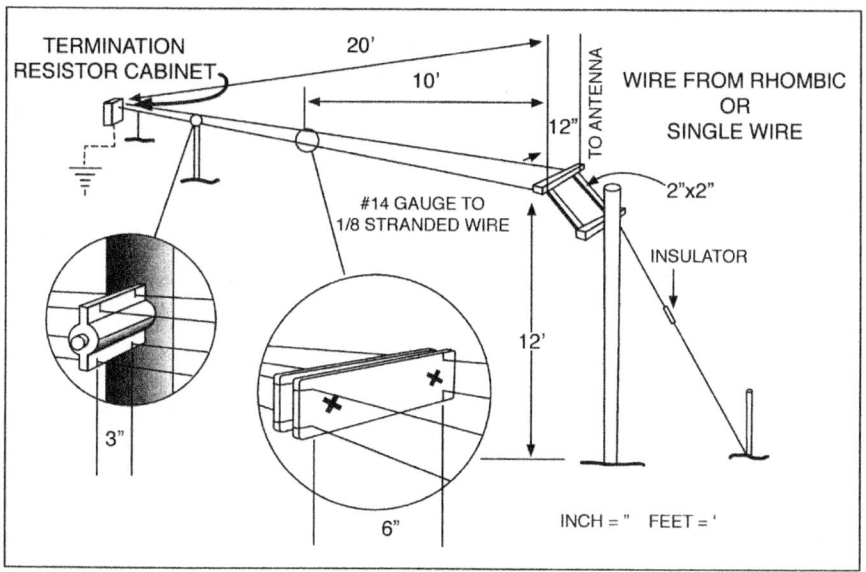

The above diagram represents any antenna system at the end of the line. The cabinet is weatherproof and may or may not be filled with mineral oil to reduce heat.

Chapter 11

Finding the Velocity K-Factor

Velocity factor is not that scary. Look at it! Visualize a long pipe that is one foot in diameter and half filled with flowing waves of water. Call this wave *current* and color it blue. As the wave goes up, it then goes down and back up again and so on. We now introduce another wave into the pipe. Call this wave *voltage* and we color it red. As the blue wave begins to go downward, the red wave starts its rise upward. As the red wave reaches the top of its wave, it now begins to drop downward as the blue wave starts upward—both keep crisscrossing at 180 degrees out of phase.

The pipe here is coax cable and the waves are RF, consisting of voltage and current, transmitted energy flowing through the coax cable toward the antenna. A transmission line has only one major function—the transfer of RF energy power from a transmitter to the antenna or from the antenna to the receiver!

K-factor, anyone?

The impedance inside a lossless transmission line repeats every lambda/2, which means if you want the impedance of the antenna to be the same inside the transmission line, then the length of the line must be lambda/2. This is also evident from the Smith Chart whose values repeat every lambda/2. The SWR is the same inside the lossless line, irrespective of its length, because it only depends on the magnitude of the reflection coefficient, and does not consider phase.

Therefore, cut coax feedline is only for single band operation in multiples of half-wavelengths with the K-factor in mind (this is the velocity factor). However, your feedline can be any length if used on a multiband antenna using tuned traps, coils, or with an antenna tuner using termination resistors at the ends of that antenna.

Remember: If the feedline is cut to half-wave in lengths (or whole multiples of a half-wavelength), the load impedance will equal the antenna impedance if the K-factor was factored in for a resonant antenna, *for a single-band operation*. This is a good thing!

A single band antenna must never be tuned out of band for any reason with an antenna tuner. Damage could occur to your tuner or plastic insulators could catch on fire, unless employed with traps, coils, and *designed for more than one band operation*.

If the feedline is any length other than the above, such as odd multiples of a quarter-waves or plain random, then the feedline will react as an impedance transformer. An impedance mismatch will be present in a single band antenna. Even if the mismatch is slight, it can be seen in the reading of the SWR. However, for *multiband operation with or without a balun*, any length of feedline is okay.

Feedline for *mono band antennas* (single-band operation) should be cut precisely for better performance.

Let's take a moment to understand the math necessary to find the velocity factor of RF cable. Many ham operators already know these equations, but for those that do not it will be explained.

Velocity Factor for Coax

Dust off the ole calculator again. Let's say you have 50 feet of coax and you want to build a feedline for a 20-meter mono-band dipole at 14.300 Mc. You must use this formula:

$$K = 984 \text{ V/V divided by the frequency}$$

The V/V is the ratio of the actual velocity to free-space velocity. Do not panic! Details are on the way. But first, it must take a look at velocity and then explain about the factor.

Velocity of propagation of radio frequency (RF) transmitted waves traveling along a wire line is nearly that of waves in free space, near that of light speed. In coax cables as with others, the velocity is materially reduced. Because of these deviations from free-space velocity, the physical length corresponding to the full-wave electrical length is less than it would be in free-space and the given factor can be found in the above formula.

Let's work out the problem above. We want to cut a proper length of coax for 14.300 Mc (at center frequency) keeping in mind the K-factor of the coax and using the formula above. Most coax velocity is .66 in the K-factor. What we must do first is to figure out how many half-wavelengths there are in our 50-foot cable.

Please listen closely! We know we have a 50-foot section of coax.

We know the frequency. First find how many half-waves at 14.300mc will fit within the 50-foot length of our cable without going over. With your calculator, enter in 468 and divide it by our frequency 14.300mc. When you're done, you will have 32.72 feet. Obviously, we only have one half-wave that can fit within the 50-feet section of the coax because 32 feet is a half-wavelength length for 14.300Mc. If we went two half-wave lengths, it would be over 50 feet and would not fit.

So, we have only one half-wavelength and we must now convert this half-wave into decimals. One-half of a wavelength is 50 percent of a wavelength; in decimals it is .50 of a wavelength.

Therefore, using the formula given we can now work out the formula.

984 X .50 X V ÷ f = L Length

X = multiply

÷ = divide

f = frequency

.50 = half-wave

V = velocity

Now multiply 984 by .50 (half-wavelength for the frequency we want). You now have 492 showing.

Now multiply 492 by (V) the velocity factor, which is .66. With 492 showing on your calculator, multiply .66 then press the equal key. The total is 324.72. If this does not compute, start over. With a total of 324.72, press divide and enter our frequency of 14.300, then press the equal key. You should read 22.7076.

You now know the length in a single half-wave multiple of 22 feet plus. Your cable is 50 feet long and now with the K-factor in play, you can see that 22.7076 is nearly twenty-three feet. It will now fit in the length of the 50-foot cable by two times. Your calculator reads 22.7076. By looking at the figures we can see that 22.7076 will go twice into 50 feet. So, with 22.7076 showing, press multiply and enter 2.

Now hit the equal key and you will show 45.4152 feet (will fit without going over). You're almost done!

Let's say it is 40 feet from your radio to the antenna outside. You do not want to use the entire 50-foot section even if it did come with connectors at both ends. Why? Current will be in conflict with voltage causing a mismatch if the full 50-foot cable is used.

However, if your antenna uses terminating resistor(s), any length of coax is fine because your antenna tuner will compensate for the difference.

In a mono band antenna (that is to say, cut for one band only) you should cut the coax using the K-factor! On the other hand, using traps, coils, or termination resistors will allow you (with an antenna tuner) to operate any band starting with the lowest frequency the antenna is cut for and all other frequencies higher.

Earlier we learned how to convert decimals into inches. You must do the same here. We have 45.4152 feet. We must continue and convert the 0.4152 into inches and then to the sixteenths. Please reread Chapter 7 if needed.

You now know where to cut the coax and re-solder a new connector. There you have it. That wasn't bad, was it?

Chapter 12

Antenna Tuners

A Tuner is a Coupling Device

An external antenna tuner should always be used when using ladder line as a feedline or using a converting coupler, which allows the change from a balance ladder line to unbalance coax. If you want to use an external antenna tuner, you must disengage your transceiver's on board auto tuner, should your radio have one installed. The reason for this is two-fold.

The first reason is most, if not all, newer radio transceivers have only a SO-239 output antenna connector installed for access to coax cable. Ladder line will not connect to the back of any radio.

The second reason is your radio tank circuit is tuned to 50 ohms and older rigs up to 75 ohms. So, the feel line or coax line must be 50 to 75 ohms with a PL-259 connector attached to the end that will connect to the radio.

As you learned earlier in this book, radio transceivers with an installed antenna tuner will only load up with a good SWR from ranges +/- 35 ohms to 150 ohms. If your antenna loads up with an SWR under 2:1, then your radio will operate simply fine. Most transceiver-installed antenna tuners will not load up antennas that display more than 130-150 ohms. To do so, an operator needs an external antenna tuner that has that required ohmage capability.

A tuner is a coupling device; therefore, an external antenna tuner should always be used when using ladder line as a feedline or using a converting coupler, which allows the change from a balance ladder line to unbalance coax. If you want to use an external antenna tuner, you must disengage your transceiver's on board auto tuner, should your radio have one installed. The reason for this is two-fold.

The first reason is, most, if not all, newer radio transceivers have only a single SO-239 output antenna connector installed for access to an antenna coax cable. Some radios will have two or three output SO-239s for multiple (coax) fed antennas. Ladder line will not connect to the back of any radio that I am aware of.

Second, your radio tank (transmitter exciter) circuit is tuned to 50 ohms and older rigs up to 75 ohms. So, the feedline or coax line must be 50 to 75 ohms with a PL-259 connector attached to the end of the coax cable that will be connected to the radio.

But, on the other hand, if your inboard tuner does not load within the aforementioned ohmage range, you will need the use of an external tuner that will load to a higher ohmage range and allowing ladder line connections.

If you desire to use an external antenna tuner, you must disengage the radio's built-in antenna tuner.

Note: For higher ohmage ranges, seek out the better-quality antenna tuners.

Not all antennas tuners will load up all wire antennas. As a rule of thumb, the more expensive the tuner, the higher ohmage (impedance) capability it becomes and the more RF power it should handle. Truth be told, the author understands why most manufacturers do not want to educate the public on the ohmage capabilities of their tuner models when they advertise.

It would be to your benefit to fully understand features governing an **antenna tuner**, such as the maximum wattage and **o**hmage a tuner will handle. Does it have a built-in dummy load? How many antenna outlets? The point is once you bring it home, will it service your needs?

You learned earlier in this book that a dipole is a balanced antenna. This means the positive wire side of the feed point (apex) is the same length as the negative side of the apex. Hence, both sides of the wire are of the same length—therefore it is balanced.

In brief, I want to add that an inverted vee is also a balanced antenna; that is to say, both wire sides are of the same length—but it is not a dipole! A dipole is designed to a specific frequency and can also become an inverted dipole by sloping the wire ends toward the ground but not touching the ground. Whereas an inverted vee are that both sides of the antenna wire are of the same random lengths measuring to the lowest frequency you wish to operate and requires an antenna tuner to load.

I hope this book will bring you to a better understand of HF wire antennas. If so, I believe you will be more refined as an Elmore to others…73's.

Jerry E. Bustin, KR7KZ

Notes

Coax or Ladder Line

All feed lines consist of coax or ladder line. Some feed lines combine ladder line and coax. You should use the simplest approach. One should always expend less effort, sweat, and tears while creating a lower SWR on their antennas. So why should we consider using ladder line when we could use coax. Ladder-lines are 300, 450 and 600 ohms, while military grade coax is 50, 75 or 100 ohms for military grades. A tank circuit of a transceiver is 50 to 75 ohms respectively speaking.

Therefore, an operator should minimize mismatch with respect to the transceiver and the antenna.

An external tuner must always be used when using ladder line as a feed line. If you want to use an external antenna tuner, you must first disengage your transceiver's auto tune (the built-in auto tuner in some transceivers). The reason for this is simple! Most HF gain radios will have SO-239 coax connector fitting and will accept only coax with a PL-259 connector attached. Ladder line will not connect directly to the back of a transceiver. Why is this? The transceiver wants to see 50 to 75 ohms. In order to use ladder line, you must connect it directly to an antenna tuner with coax from the antenna tuner to back of the ham radio.

Most built-in antenna tuners will operate up to or slightly past 125 ohms (presently known). Your transceiver's built-in antenna tuner may not load up (good SWR) on an antenna consisting of 150 ohms or more. To do so you must have an external antenna tuner capable of covering antenna ohmages out of this impedance range. You can use ladder line or coax cable with or without a balun or unbalun.

A dipole with an unbalanced feed line such as coaxial cable is used for transmitting. The shield side of the cable radiates. This can introduce RF currents into your shack causing other problems to your electronic equipment. The coax cable inside your shack will radiate an RF field, causing RFI interference. That is why if you decide to use coax cable as your feel line, as most hams do, always use a Balun at the feed point.

Not only will the RFI of unwanted current stop from running down the ground braid of the coax cable and into your shack, but it will also distribute the output current into the dipole evenly. For the high-power folks who love their amplifiers, the author suggests installing a high wattage choke isolator in the coax line about fifteen to twenty feet away from the dipole for the same reasons.

In this book, any drawing showing coax as a feed line may be substituted with ladder line or vice versa—if the ladder line ohmage is close to the antenna's impedance (ohmage) and connected to an external antenna tuner that is capable of handling the impedance of your antenna.

Note: Not all antenna tuners will load up all wire antennas. As a rule of thumb, the more expensive the tuner, the higher the (ohmage) impedance capability and the more RF power it will handle. The author wishes the manufacturers would display the ohmage capability of their tuner models in their ads. One should always consult the manufacturer before you buy!

Author's Thoughts on Gain

Ham operators seem to have two differing opinions when considering gain. On one side of the fence, you have those that will standby a reference antenna as only being an isotropic radiator. On the other side, you'll have justified disagreement. The author stands with the "other side."

From the minds of ole timer wisdom, the author offers some educated discussion concerning gain.

"In antenna theory, one may use as a reference antenna either a 'dipole' or an 'isotropic radiator'. The dipole has an undisputed TRUE gain of zero db. Yet it is obscured in its directivity. Whereas an isotropic radiator is of a 'pure imaginary theoretical radiator'; having a directivity of zero dBi/dB that the radiator will equally transmits and receives electromagnetic radiation from any arbitrary direction. However, please understand in reality a coherent isotropic radiator can never exist and a dipole is not coherently pure." *(Reference: Isotropic radiator…Helmholtz's Wave Equations (a dipole is at 1.76 dBi), Maxwell's/Hertzian's Equations (a dipole is 2.15 dBi), and in the U.S. Military training manual, a dipole is 1.5 dBi over an isotropic radiator. Take your pick.)*

Read the manufactures information before you purchase. If they quote gain from an isotropic measurement, it sounds great. However, knowing your theory of dipole gain will help you understand how you can be misled.

Good Source for Information about Amateur Radio
- http://www.twit.tv/hn
- HamNation is a great ham radio weekly podcast
- (The players at the podcast studio are at the top of their profession in ham radio and a high credit to Amateur Radio Services. Consult Wednesday's listings.)

ABOUT THE AUTHOR

Jerry E. Bustin was born in Phoenix, Arizona. A relative recalled, "When Jerry was eight years old, I saw him soldering. I asked him what he was doing? He said I'm going to find out how this radio works!"

In his late teens, Jerry joined the Army and later taught technical classes. He was sent to Vietnam twice where he served our nation honorable as a crew chief/door gunner on *UH-B Huey* gunship. He was decorated several times. The rest of his time in the service was in MARS radio stations throughout the world.

Jerry's ability to explain technical subjects to novice and journeyman alike is unmatched. Making difficult subjects seem simple is a talent few people possess. However, when your heart's in what you teach, it is possible, as the author has demonstrated.

Printed in Dunstable, United Kingdom